NATO's Conventional Defences

Options for the Central Region

Stephen J. Flanagan

MACMILLAN
PRESS

in association with the
Palgrave Macmillan

First published 1988

Published by
THE MACMILLAN PRESS LTD
Houndmills, Basingstoke, Hampshire RG21 2XS
and London
Companies and representatives
throughout the world

British Library Cataloguing in Publication Data
Flanagan, Stephen J.
NATO's conventional defences: options for
the Central Region.—(Studies in
international security; 27).
1. North Atlantic Treaty Organization
2. Deterrence (Strategy)   3. Warfare,
Conventional
I. Title   II. International Institute for
Strategic Studies III. Series
355'.0217   UA646.3
ISBN 978-0-333-46368-0   ISBN 978-1-349-19484-1 (eBook)
DOI 10.1007/978-1-349-19484-1

Published by Palgrave Macmillan in association with the International Institute for Strategic Studies

Studies in International Security

International Institute for Strategic Studies conference papers

Christoph Bertram (editor)

Robert O'Neill (editor)

For Lynn Wansley Flanagan and our sons
Brian and Neil.
May they never know war.

Contents

Preface

Since the early 1980s a seemingly endless array of proposals for improving NATO's conventional military posture in Central Europe has surfaced, leaving not only the attentive public, but many officials and experts as well, quite bewildered. Most of these concepts have been advanced in a disjointed fashion, but with the common goal of reducing Allied dependence on nuclear weapons. Proponents have generally underspecified how their new ideas would serve the overall objectives of Western strategy. This book reviews the full range of recent official and non-official proposals for improving NATO's conventional posture, from exploitation of emerging technologies to non-provocative defences, in the light of prevailing military, political, economic and demographic trends that will influence their fate. The book goes on to provide a framework for assessing which options should be pursued in the light of various possible strategic objectives for the non-nuclear component of Western deterrence. It concludes with an assessment of which strategic concept for conventional defence is both militarily most desirable and politically sustainable.

The shape of NATO's conventional capabilities achieved renewed salience over the past decade as a consequence of the acrimonious debate over Allied nuclear weapons and arms control policies and widening transatlantic differences on a number of other issues. The attraction of a deterrent posture less reliant on nuclear weapons lingers on both sides of the Atlantic, despite considerable divergence of views over the structure and most appropriate means to achieve such a posture. Yet, in the aftermath of the nuclear abolitionist sentiments expressed by both President Reagan and General Secretary Gorbachev at Reykjavik, there are also fears, particularly strong in Europe, that conventional deterrence has failed too often in the past ever to be relied upon as the primary guarantor of Western security. Moreover, the political mood and economic and demographic trends throughout the Alliance are not propitious for an expansion of conventional capabilities. Thus a vigorous debate has been reopened on the enduring questions of what mix of nuclear and conventional forces can best ensure peace and how best to manage available defence resources.

In the midst of this debate over military strategy and resource

viii

allocation, European discontent with the direction of American policy has revived interest in the development of a distinct defence identity through more extensive political consultation and military cooperation. Significantly expanded European cooperation in defence planning and armaments production could result in a more effective utilization of defence resources and a strengthening of the political underpinnings of the second pillar of the Alliance. Indeed, genuine progress in this area will be critical to dampening mounting domestic sentiment in the US favouring reduction of the American military presence in Europe. However, given this context of enduring European and American differences over the nature of and appropriate means for dealing with various security problems, a prudent transatlantic dialogue and careful management of these problems will be essential to the avoidance of further polarization of the Alliance.

This book has been informed by discussions and interviews with a broad range of officials and independent analysts in Europe and North America. A number of officials were most generous with their time and regrettably their assistance cannot be acknowledged more explicitly. A number of colleagues, including Stephen Biddle, Kurt Campbell, Ivo Daalder, Eckhard Lübkemeier, John Mearsheimer, Andrew Pierre, Barry Posen, Christian Tuschhoff, Gregory Treverton and Stephen VanEvera have been particularly helpful in commenting on all or part of earlier drafts of this manuscript. This book has also benefited considerably from a rigorous review by the Directing Staff at the International Institute for Strategic Studies (IISS). The late Deputy Director of the Institute, Colonel Jonathan Alford, was my original mentor in this project, helping me shape a coherent plan of attack and offering sound editorial judgement. It is with a deep sense of regret that I can only pay honour to his memory here, and not thank him directly. Along the way, the Institute's former Director, Robert O'Neill, Acting Deputy Director Kenneth Hunt, and former Directors of Studies Lynn Davis and Robert Nurick provided sound advice and firm encouragement. Finally, I have been most fortunate to have as the manuscript's final godfather the current Deputy Director of the Institute, Colonel John Cross. The book now reflects the benefits of his patient and painstaking review of the final few drafts. Colonel Cross' knowledge of military detail, sound grasp of the big picture and keen editorial eye helped me fill several analytic gaps and reconcile differences in the common language we Americans share with our British cousins.

This book was begun while I was an International Affairs Fellow of the Council on Foreign Relations and a Research Associate of the IISS. I am deeply indebted to the Council, particularly the Directors of the fellowship programme, Alton Frye and Kempton Dunn, and to its sponsor, the Ford Foundation, for their support. My period of residence in London was made possible by a grant to the IISS from Stiftung Volkswagenwerk. Barry Blechman was kind enough to offer me encouragement and a temporary US base at the Center for Strategic and International Studies in Washington. Finally, this book was completed during my years at the Center for Science and International Affairs in Harvard's John F. Kennedy School of Government. It would be impossible to catalogue the valuable insights I received from my interactions with so many fine colleagues in Cambridge, Massachusetts. However, I do want to thank the Center's current and former directors, Joseph S. Nye, Jr, and Paul Doty for their support. Finally, much credit is due to Marie Allitto-Hadley, Valerie Grasso, Veronica McClure, and Mary Ann Wells at Harvard and Helen Rayner at the IISS for ensuring a very smooth production of the several draft manuscripts.

Despite all this fraternal assistance, I alone must take full responsibility for any shortcomings in the final product. Throughout all of this undertaking, my wife and two sons have sustained my morale and forgiven long absences. It is therefore most appropriate that this book is dedicated to them.

STEPHEN J. FLANAGAN

1 Introduction: The Search for a Stable Deterrent

A general consensus has emerged in the West over the past few years that NATO's conventional military capabilities should be increased as a way of reducing what many regard as excessive – and unrealistic – Allied dependence on nuclear weapons in deterring Soviet aggression in Europe. The October 1986 Reykjavik Summit, the INF accord and other proposals for dramatic reductions in, or even total elimination of, various categories of nuclear weapons in Europe has heightened concern about the role and adequacy of NATO's conventional defences. Growing doubts about the credibility of the US nuclear guarantee and durability of the US military presence on the Continent have also stimulated West European interest in expanded defence cooperation, both within and outside the NATO framework. However, views among the Allies and within member states diverge sharply over the specific conventional force enhancements that should be pursued and the military plans, doctrines and tactics they should serve. There are also considerable differences throughout the Alliance over the course Western arms control policy should pursue and its place within the spectrum of interstate relations.

The widespread anxiety about the extent of NATO's reliance on nuclear deterrence that re-emerged in the late 1970s served as a catalyst for this burgeoning interest in conventional forces. The threatened deployment of US enhanced radiation warheads (ERW) on short-range systems sparked considerable outcry among European publics during the 1977–8 period.[1] The nuclear allergy was exacerbated by mounting doubts about the safety of nuclear power plants. However, it was the December 1979 'dual-track' decision on modernization of intermediate-range nuclear forces (INF) that triggered NATO's most recent nuclear crisis. While NATO has weathered a number of crises over the past three decades, none has shaken the Alliance so profoundly as this enduring turmoil over the role of nuclear weapons in Western strategy.

This general unease about nuclear matters was compounded by fears about the collapse of East–West *détente* after the Soviet invasion of Afghanistan. The American shift to a more confrontational posture towards Moscow, remarks by senior Reagan Administration officials implying the possibility of limited nuclear war in Europe, and Washington's decision to proceed with production of the ERW further heightened Europe's nuclear jitters. By the time the Administration announced its willingness to reopen the INF negotiations in November 1981, anti-nuclear protest in Europe was attracting increasingly broad support.

In addition to this public uproar, new doubts about the credibility of NATO doctrine and the American nuclear guarantee surfaced. Many prominent observers and officials took the view that the erosion of American superiority in strategic and theatre nuclear systems, coupled with the impressive growth of Soviet conventional capabilities, required a reassessment of NATO strategy.[2] It was argued that NATO's military standing was so bad *vis-à-vis* the Warsaw Pact that it would almost certainly have to resort to early use of tactical nuclear weapons in a military conflict. As US Senator Sam Nunn commented in 1977, the growth of Soviet capabilities had changed NATO strategy from flexible response to 'inflexible response'.[3] Nunn and others contended that qualitative improvements during the 1970s in the Warsaw Pact's numerically superior general purpose forces also called into question NATO's ability to sustain a forward defence. Thus the overall deterrent value of NATO's military capabilities was perceived to be declining. Senator Nunn, Supreme Allied Commander Europe (SACEUR), General Bernard Rogers, and other US and European authorities voiced with renewed urgency their long-standing conviction of the need to bolster NATO's conventional forces.

These preliminary exhortations for shoring up NATO's conventional posture were advanced within the context of maintaining the strategy of Flexible Response. However, participants in this most recent rebellion against NATO's nuclear predicament range from those who affirm the deterrent value of the full range of NATO's nuclear weapons but want to reduce the probability of early first-use, to those who seek to remove all nuclear weapons from Europe and develop a purely conventional defence.

In early 1982, four prominent former US government officials reviewed this military situation and concluded that NATO should renounce the option of using nuclear weapons first in response to Soviet aggression.[4] The so-called 'Gang of Four' argued that adoption of a no-first-use posture would also spur improvements in NATO's conventional capabilities because such a policy would not be viable without more capable general purpose forces. The 'Gang of Four' proposal found few supporters in Europe.[5] European critics of no-first-use noted how the important deterrent value of uncertainty regarding the nature or scope of Western military responses would be lost with such a doctrine, and that it would not be credible anyway without a removal of NATO nuclear weapons.

The initial deployment of 108 Pershing II and 128 ground-launched cruise missiles (GLCM) by the end of 1986 marked what seemed to be the end of the INF challenge to Atlantic solidarity, since this action was perceived as a demonstration of deep-seated Alliance resolve. In the aftermath of these deployments, the Alliance might have been expected to settle into a period of apparent tranquillity. However, while nuclear protesters are no longer taking to the streets of European capitals, lingering discord over nuclear policy threatens NATO's unity in the long term. Alliance solidarity on nuclear policy has encountered several new challenges.

Most Europeans abhor the nuclear war fighting implications of the US Strategic Defense Initiative (SDI) and worry that Washington will press ahead with this programme. Many Europeans feel that the possible development of a defensive shield over the US increases the possibility of a nuclear war limited to Europe. As in the past, there has been no fundamental resolution of the role of nuclear weapons in Allied strategy, only a sublimation of disagreements. President Reagan's condemnation of Mutual Assured Destruction and his pledge to make nuclear weapons 'impotent and obsolete' by developing effective defences have actually revived support for nuclear deterrence in Western Europe. Pursuit of SDI undermines the concept of shared European and American risk that is a fundamental component of Western deterrence.[6]

At the same time, Gorbachev's dramatic arms control proposals of early 1987 rekindled hope that the nuclear build-up in Europe could be reversed. However, most European officials remained chary of arms control proposals which would remove theatre-based American nuclear weapons from the military equation.[7]

Most Europeans have no interest in relinquishing the ultimate assurance afforded them by the US nuclear guarantee; rather, they want to achieve the most stable mix of nuclear and non-nuclear means possible to enhance their security. The vast majority of military planners and opinion leaders continues to support current strategy, but is uncomfortable with NATO's military posture because it does not provide high confidence that the Alliance could avoid resort to relatively early first-use of nuclear weapons to halt an offensive by superior Warsaw Pact ground forces. Thus the current debate in official circles is largely about making existing strategy and doctrine work rather than searching for radically different options.

None the less, there has been increasing discussion of the possibility of radical shifts in military postures, some of which would require alterations in NATO doctrine. The post-Reykjavik revival of nuclear abolitionism and other proposals for very deep cuts in superpower nuclear arsenals would threaten NATO's Flexible Response strategy if they were put into practice. There is growing interest in expanding European defence cooperation, both as a way to strengthen this pillar of NATO's defence structure and as a hedge against American disengagement. Some West European political parties, including the West German Social Democratic and British Labour parties, go further and have endorsed radical alterations in force structure and doctrine.

The search for an improved conventional posture will be hampered by political and economic trends in the NATO countries which militate, as they have for the past 30 years or more, against the significant increases in defence spending and expanded military cooperation that would be required. Indeed, fiscal and manpower constraints in all NATO countries will limit their capacity even to maintain current military postures, let alone expand their conventional forces.

More fundamentally, many West Europeans find a military strategy which envisages a protracted conventional war on the Continent an unacceptable 'cure' for nuclear *angst*. Indeed, such a remedy would only revive fears of a decoupling of European security from the American nuclear guarantee and of a rise in Soviet adventurism as a consequence of reduced risks of escalation. Similar concerns have been aroused by Washington's contemplated development of SDI, which might provide America with some defence against inter-continental ballistic missiles (ICBM) but leave Europe vulnerable to a wide array of other threats.

In addition, the European members of NATO are certain to show some resistance to any new strategy or politically-charged conventional force modernization plan that appears to be imposed on them by Washington, particularly if commitment to such efforts is transformed into yet another test of Alliance solidarity. Manifestations of such renewed resistance have already surfaced. The Nunn Amendment, introduced in the US Senate during the summer of 1984 and likely to return in some form, called for phased reductions of American troops in Europe unless the West European members of the Alliance undertook certain conventional force improvements.[8] It is likely that the Reagan Administration's successors will be forced by political or economic imperatives to cut defence spending. Many West Europeans worry that the expensive presence of US troops in their countries remains a most attractive target for reductions among American politicians who believe that European members of the Alliance should spend more on their own defence. Furthermore, various – largely American – concepts for strengthening or enlarging the non-nuclear component of NATO's deterrent through the introduction of innovative tactics and emerging weapons technologies have received an extremely cool reception in Europe.[9] These concerns illustrate the potential for transatlantic disharmony attendant on the conventional defence debate.

Whatever strategic concept ultimately guides NATO defence planning in the coming years, it appears most likely that only incremental improvements in NATO's conventional capabilities will be achievable. Some of these improvements would clearly be stabilizing and can assuredly be achieved by means of a modestly improved allocation of resources, better defence planning, selected application of new weapons technologies, and by the adoption of certain tactical innovations. Implementation of such initiatives will require judicious consultation both among Allied governments and within member states. Without such consultations, disharmony over the ways, means and goals of improving NATO's conventional posture could become a major source of tension within the Alliance during the coming decade.

Within this context, this book reviews the principal conventional force improvement proposals which have surfaced in recent years in the light of the general political, military, economic and demographic trends that are likely to determine their fate. But unlike other studies of these issues, it advances guidelines for assessing NATO's military options in the context of various strategic concepts for conventional defence.[10] It also suggests which defence programme and planning mechanisms might enable the Alliance to sustain an effective, credible and affordable deterrence posture in the changing strategic environment of the late 1980s and 1990s.

The book begins with a survey of past NATO conventional force modernization plans in Chapter 2, where the implications of these experiences and of the current state of affairs within Alliance countries for conventional force improvements are also addressed. Chapter 3 examines the prospects for expanded cooperation among all NATO countries in armaments development and among the European members of the Alliance in defence planning and operations. The changing role of France and the new presence of Spain in the Western Alliance, as well as Franco–German defence cooperation, are reviewed in Chapter 4. These developments could greatly enhance confidence in NATO's conventional defence posture, particularly if France altered its forces and infrastructure in ways that made its commitment to the defence of Germany more manifest and credible. Chapter 5 assesses the likely impact of emerging non-nuclear weapons technologies on the conventional balance, and attendant economic and political considerations that will influence their development. Chapter 6 explores the effectiveness of several military proposals including new operational concepts, such as the Follow-On Forces Attack (FOFA) plan, and adjustments to the current roles and missions of Allied forces in order to improve NATO's deterrent posture and war-fighting capabilities.

Chapter 7 considers the desirability of various conventional force improvement initiatives in the context of several strategic concepts which could guide NATO defence planning. These concepts include: the current NATO policy of planning for a short conventional phase before escalating; improving conventional forces to prolong this conventional phase; developing capabilities to pursue a more protracted conventional war, ranging from 40 to 60 days – as reportedly called for in the current US Defense Guidance – to sustaining non-nuclear war indefinitely or as long as the Warsaw Pact does; the introduction of an offensive, counter-attack strategy, as advocated by Samuel Huntington and others; and the adoption of a totally non-provocative, defensive military posture along the lines advocated by various European strategists.

The challenges confronting NATO leaders in the development of effective arms control initiatives and of responses to likely Soviet military and diplomatic actions are considered in Chapter 8. Finally, the last chapter explains why extending somewhat the length of time NATO could wage a conventional war before escalating to nuclear use is the most prudent strategic concept for the West. It examines the most appropriate conventional force enhancements for supporting this concept and the political and economic impediments that could inhibit their realization. Chapter 9 also advances a number of practical policy options and institutional mechanisms which could be useful in enhancing the effectiveness and cohesion of NATO conventional defence plans, and suggests strategies for reconstructing a consensus on defence policy within the Alliance.

2 The Elusive Consensus

The prospects for realizing dramatic improvements in NATO's conventional capabilities do not appear auspicious from the perspective of enduring political and emerging economic and demographic trends. The main issues in the contempoary conventional defence debate echo concerns first heard 20 years ago. As in the past, there is little interest on either side of the Atlantic in expanding the size of conventional forces, and the dwindling manpower pool may preclude such a move, yet adding more brigades to Allied armies would be one of the best ways to improve conventional deterrence. Restraints on governmental spending throughout the Alliance will almost certainly limit the amount of resources available for military improvements, and will require difficult trade-offs to be made in formulating defence budgets. In developing its conventional defences, the Alliance will also have to confront new challenges from the Warsaw Pact on both the political and military fronts. While the East will also have to cope with resource constraints and demographic changes, Warsaw Pact military forces are likely at least to keep pace with NATO in their sophistication, effectiveness and size.

This chapter describes the context of the conventional defence debate. It outlines the historical evolution of this debate and considers the contemporary implications of the experience. It goes on to examine how political, economic and demographic trends in the West will affect conventional defence plans.

HISTORICAL TRENDS

Despite periodic concern about overreliance on nuclear weapons, NATO countries have consciously opted for the less costly deterrence that these systems provide instead of developing more robust conventional defences. The non-nuclear component of Western deterrence has experienced cyclical periods of benign neglect and fleeting alarm. Allied concern about the conventional balance in the 1950s was soon allayed by the new nuclear guarantee. American interest during the 1960s in developing a largely conventional deterrence encountered little enthusiasm in Europe, and US attention was soon diverted to the war in South-East Asia. The Soviet

8

achievement of nuclear parity with the US in the 1970s stimulated yet another re-examination of the adequacy of NATO's general purpose forces and many unmet pledges to improve shortcomings. Most recently, the INF agreement and further proposals for reducing – or even eliminating – nuclear weapons in Europe have refocused attention on the state of the conventional balance and triggered a debate on the military and political initiatives NATO would have to pursue in order to maintain stability if these proposals were implemented. Throughout these periods, NATO general purpose forces have grown little in size and sustainability, but markedly in mobility and firepower.

From Lisbon to the Seductive 'New Look'

The basic structure of post-war Western security arrangements was realized with the ratification of the North Atlantic Treaty in 1949. However, several issues had to be addressed before this structure could take coherent military shape. The most important of these unresolved issues concerned the nature and size of the US and West European military contributions and the role of West Germany in this new security partnership. The Korean War galvanized domestic support for enlarging the US troop presence in Europe to thwart what was seen as the global challenge of communism. In September 1950, President Truman announced a significant increase in the size of American forces stationed in Europe, knowing that he would have to convince the French of the benefits of German rearmament in order to secure Congressional support for the maintenance of this military commitment.[1] Many conservatives in the Congress felt that the West Europeans should assume the leadership and the greater part of the burden of their own defence before the US made any long-lasting commitments.

In mid-1950, French Premier René Pleven proposed the establishment of a European Defence Community (EDC), essentially a European army within which token West German units would be incorporated. The French offered this concept as a way to slow the pace of German rearmament and to secure for themselves a leadership role in the development of Western security plans. At the same time, Paris hoped to convince Washington of the need to commit itself to forward defence of its Allies by deploying sizeable forces in West Germany. At their December 1950 ministerial meeting, the North Atlantic Council (NAC) approved a compromise measure which endorsed French plans for creating a European Defence Force with the condition that the US should begin deploying units to Europe and that West German troops should become available for Western defence under strong provisional controls until a permanent system of European cooperation could be developed. As Timothy Ireland has characterized this deal, the US won French endorsement for the idea of German rearmament, and the French gained an immediate US commitment to the defence of Europe while delaying German rearmament.[2] At this session the ministers also ratified the establishment of a Supreme Allied Headquarters, with an American commander.

At their February 1952 meeting in Lisbon, NATO foreign, defence and finance ministers adopted 'firm force goals' of 50 active divisions, 46 reserve divisions, 4000 aircraft and strong naval forces by the end of 1952, with further increases to be effected during the subsequent two years. The ministers balked, however, at endorsing the recommendations of military planners who suggested that 96 active divisions and over 9000 aircraft would be necessary to defend adequately against the 175 smaller Soviet divisions then estimated to be available for an attack on Western Europe.[3] At Lisbon, the ministers also finalized the principles governing the relationship between the EDC and NATO. Three months later, Washington spurred the Allies to sign the EDC treaty and the agreement between the Federal Republic and the three occupying powers re-establishing German sovereignty.

It soon became evident that sustaining peacetime military capabilities of this magnitude in the West would be impossible because of economic and political constraints. British and French defence budgets, limited by the costs of post-war economic recovery, were already stretched by overseas military commitments.[4] Given the growing manpower demands of French operations in Indo-China and the multitude of British deployments outside Europe, neither government could realistically expect to expand its European-based forces. The Eisenhower Administration entered office in 1953 with a fiscal austerity programme, and hoped that further cooperation among the European allies under the aegis of the EDC would enable the US to reduce the cost of its Continental commitments. Thus the US government, which had initially been sceptical of the plan's feasibility, became the most enthusiastic supporter of the EDC as the best way to foster European unity and involve Germany in Western defence.

US hopes for enhanced European cooperation proved, however, to be ill-founded. Ratification of the EDC Treaty, which allowed greater German rearmament than envisaged in the Pleven Plan, foundered in the prevailing turmoil of French politics. In the meantime, the Eisenhower Administration had already initiated plans to increase the role of strategic and tactical nuclear weapons in NATO's overall strategy. This nuclear shift worked at cross-purposes with the original idea of the EDC. It implied, particularly to the French, that the US doubted the feasibility of developing a robust conventional posture against the USSR, even if West German forces were incorporated into Western defences under the EDC. Moreover, the French feared that once the EDC was in place, it would allow Washington to bring its troops home from Europe. The French National Assembly's ultimate rejection of the EDC Treaty in August 1954 turned primarily on lingering fears of German resurgence at a time of uncertainty at home.[5] This setback was mitigated somewhat by the signing two months later of the Paris and London Agreements which, together with the protocols of the West European Union (WEU), made possible West German entry into NATO. None the less an important opportunity for forging a stronger conventional defence posture and a more integrated European pillar of the Alliance was thus lost.

With the collapse of the EDC, the US moved quickly to integrate West Germany into Allied defence by other means. The sovereignty of the Federal Republic was recognized in October 1954, and Bonn agreed to allow the stationing of foreign troops in numbers at least equal to the then current levels of occupation forces in exchange for a formal end of the post-war occupation. Forward defence to maintain the integrity of all West German territory became, at West Germany's insistence, another central tenet of NATO strategy.

These experiences led many NATO planners to conclude that the West would simply have to find some way to cope with conventional inferiority. The Eisenhower Administration's 'New Look' philosophy seemed to offer a relatively cheap solution to this problem. Eisenhower's organizing principle was that American security rested on 'two pillars': military strength and a vibrant economy. In Eisenhower's view a new balance between a strong economy and a strong military capability – one with a decided tilt towards the former – had to be established and maintained over the long term.[6] Domestic opposition to the Korean War had convinced Eisenhower that the US should avoid future involvement in protracted conventional conflicts and that it could not afford to maintain troops all over the globe to provide direct local defence of vital Western interests. This new American approach to security, which stressed technology over manpower and nuclear weapons over conventional, led to adoption of the doctrine of 'massive retaliation' with nuclear weapons to be used in response to any Soviet military aggression. Deployment of considerable numbers of US tactical nuclear weapons began in 1953. At the April 1954 NATO Foreign Ministers meeting, US Secretary of State John Foster Dulles announced to his counterparts that:

> The United States considers that the ability to use atomic weapons is essential for the defense of the NATO area in the face of the present threat . . . In short, such weapons must now be treated as in fact having become 'conventional'.[7]

The European members of the Alliance were not totally comfortable with this new American policy, but they had little alternative. At their December 1954 meeting NATO Ministers therefore endorsed the concept of a massive retaliation strategy, which was formally adopted as NATO Military Committee document 14/2 (MC 14/2) of 1956.

Even as the new strategy was being implemented, some of its fundamental assumptions were overtaken by events. The New Look was founded on the maintenance of overwhelming American superiority in nuclear-capable airpower. By the mid-1950s it was clear that both the US and the USSR would develop secure strategic retaliatory capabilities, making it increasingly unlikely that either state would use nuclear forces to repulse a small attack. Moreover, military exercises at the time suggested that casualties on the 'atomic battlefield' would be so high that an effective defence in the nuclear age might actually require more rather than fewer troops. The impending growth of Soviet strategic nuclear capabilities forced the Eisenhower Administration to adopt the 'New New Look'. Rather than assuming strategic superiority and conventional inferiority, the Administration advocated the development of military sufficiency in both areas. In an era of mutual deterrence at the strategic level, it was important for the US to develop the capability to fight limited wars.

It might have been expected that this New New Look would require an increase in conventional capabilities. However, the Administration was unwilling to bear the fiscal burdens of the requisite build-up. Many US military planners found the growing arsenal of battlefield nuclear weapons attractive substitutes for manpower in repulsing limited aggression. Conventional defences were increasingly seen as useful primarily for providing a very brief conventional phase of war before a wider nuclear response. One of the most vigorous proponents of the nuclearization of the US posture was the then Chairman of the Joint Chiefs of Staff (JCS), Admiral Arthur Radford. Responding to the Administration's budget constraints, Radford proposed in July 1956 that the only way to stabilize the level of military expenditures was to reduce US armed forces by 800 000 to two million men over the following three years, with a 450 000 cut in the Army alone. The Radford Plan sought to balance these cuts in general purpose forces by maintaining superiority in strategic nuclear forces. Radford advocated retention of only token atomic task forces in Europe and Asia, to be supplemented by Allied forces, for the conduct of limited wars. [8]

Details of the secret Radford Plan were soon leaked to the *New York Times*, causing great anxiety among the European Allies, particularly the West Germans. The Adenauer government promptly dispatched the chief of the *Bundeswehr* to Washington to elaborate its conviction that strong conventional ground forces were a critical element of Western strategic deterrence. These Allied objections, coupled with significant domestic opposition, forced withdrawal of the Radford Plan. Congress shared Admiral Radford's belief in the importance of maintaining superiority in strategic airpower, but preferred to acquire this capability by enlarging the budget rather than by scaling back conventional forces. The Administration chose a middle ground, asserting that drastic cuts in conventional forces were unwise and that US strategic sufficiency could be maintained within programmed resource constraints.

Thus by the late 1950s the structure of NATO's deterrent posture had shifted to fundamental reliance on decisive retaliation with strategic nuclear weapons. Under the New New Look, the principal purpose of ground forces in Western Europe, now equipped with thousands of nuclear weapons, was to provide a 'tripwire' of limited flexibility and to hold an attack until the full weight of strategic retaliatory forces could be applied. The large US military deployments in Europe and this strategy were intertwined because the credibility of Washington's threat to use nuclear weapons in response to a Soviet attack hinged on the maintenance of this presence on the Continent. This nuclear guarantee, coupled with lessened fears of

Soviet aggression towards Western Europe in the mid-1950s, further undercut Allied interest in preparing for a prolonged conventional conflict.

NATO ministers settled into this new posture with equanimity. The NAC agreed in December 1956 that future military plans be based on the concept of forward defence but take into account the growth in Soviet capabilities and the new nuclear weapons available to the Alliance. Based on this guidance, Allied military planners developed a five-year projection, which became known as the MC-70 plan, of the minimum forces required to fulfil their missions. The MC-70 plan called for the maintenance of only 30 active divisions, the same levels recommended in the 1955 Allied 'force goals' document. Yet meeting even this goal proved elusive. It became apparent that NATO planners would have to live with existing force levels, which totalled fewer than 25 effective fighting divisions.[9] According to Western estimates at the time, these forces confronted 28–30 Soviet divisions in Eastern Europe, including roughly 400 000 troops in East Germany and 100 000 elsewhere in the region, an additional 60–70 Soviet divisions in the Western military districts of the USSR, and 1.5 million East European soldiers, of whom about half had significant combat capability. Soviet ground forces were supported by more than 10 000 tactical aircraft, roughly one-quarter of which were based in Eastern Europe.

The Origins of Flexible Response

Doubts about the prudence and viability of massive retaliation surfaced almost immediately. William Kaufmann and other civilian defence analysts questioned the credibility of Dulles' strategy in the light of the historical reluctance of US leaders to consider the use of nuclear weapons, a reluctance which seemed certain to increase as the Soviet strategic arsenal grew.[10] Similarly, many French officials argued that the credibility of the American extended deterrent would evaporate once the US heartland became vulnerable to nuclear attack. By 1957 even Dulles admitted that the threat of massive retaliation to local aggression was not very credible and suggested that battlefield nuclear weapons could provide a more effective deterrent to limited aggression.[11] Studies initiated early in the Kennedy Administration revealed several fundamental flaws in the Dulles strategy.

New threat assessments found that US military planners had overstated the Warsaw Pact threat and underestimated the strength of NATO's conventional military capabilities.[12] These revised estimates, which used more sophisticated analytic techniques and benefited from improved intelligence data, found that the balance in Central Europe looked much better for NATO than it had previously. In 1963 the US Defense Department projected that NATO had 1.5 million men under arms in Europe, compared to 2.2 million Soviet and other Warsaw Pact soldiers in Eastern Europe. The division tally also looked better for NATO. There were estimated to be 22–26 combat-ready Soviet divisions in Eastern Europe opposing 25 such NATO divisions in the Central Region. Also figuring in the Central Region were 35 divisions, of lesser readiness and dubious reliability, from other Warsaw Pact states. The Soviet Union had 60 divisions of varying capabilities in the Western Military Districts of the USSR, while NATO had another 29 divisions in Southern Europe and Turkey. Kennedy Administration officials also found that both the numbers and capabilities of Soviet tactical fighters had been greatly overstated. The air balance seemed less daunting now that the Pentagon estimated that NATO's 3500 tactical aircraft based in Western Europe faced only 1500 Soviet and 2500 other Warsaw Pact aircraft in Eastern Europe. Western air defences were also assessed to be superior to those deployed in Eastern Europe. NATO's capabilities to wage a non-nuclear defence seemed more impressive. American officials now believed that the Warsaw Pact could not muster the 3-1 advantage in forces generally judged to be required for a successful ground offensive against NATO. The Kennedy Administration did not dismiss the threat; indeed, it tried to focus attention on critical operational problems, particularly reducing NATO's vulnerability to surprise attack and improving Allied capabilities to reinforce first line forces.

More importantly, other Kennedy Administration studies brought into focus considerable uncertainties about NATO's capacity to use nuclear weapons in support of its objectives during a military confrontation. On the basis of these analyses, the Kennedy Administration presented the NATO Allies with a new strategy of 'Flexible Response' for consideration at the May 1962 Athens ministerial meetings. As envisaged by its American proponents, the new strategy would raise the nuclear threshold significantly by implementing modest improvements to conventional forces and responding to initial Warsaw Pact attacks with them alone. Indeed, as former

Defense Secretary McNamara has noted, the flexible response concept he proposed envisaged that NATO's conventional capabilities could be improved to the extent that actual employment of nuclear weapons would become either unnecessary or very limited, and not required until a fairly late point in the conflict.[13]

The West European Allies believed that the primary deterrent to Soviet aggression was the danger that war in Europe could escalate rapidly to the point where Soviet and American territory would be threatened. Their governments expressed grave concerns that the new strategy's emphasis on conventional defence would undermine NATO's nuclear deterrent. The Allies did not share McNamara's view that the build-up of conventional forces essential to Flexible Response would not require unacceptable political or economic sacrifices. Moreover, Washington's partners feared that Flexible Response was an effort by the US to decouple its security from their defence. That is to say, Europeans worried that the ulterior motive of this doctrinal shift was to prevent any conflict from escalating to the level that would put the North American continent at risk. The Europeans contended that any effort to remove this coupling might pave the way for Soviet adventurism. They also rejected categorically the notion of limited nuclear war consisting of strikes against military targets alone. Europeans were convinced that the advent of survivable retaliatory forces had created a stable situation of mutual societal vulnerability. In addition to these general concerns, the major West European governments had some particular objections to Flexible Response.

The West German governments of the time found the new emphasis on conventional responses inimical to their objective of forward defence: that is, of repulsing any attack as close to the inner German border as possible. While opposition defence expert and future Chanellor Helmut Schmidt and some members of the *Bundeswehr* welcomed the development of a conventional option, German policy reflected the ruling Christian Democrats' view that greater predictability in the early stages of NATO's defence efforts would weaken deterrence. Rather, they favoured a deterrent concept based on continuing to pose incalculable risks to any aggressor, including an explicit risk of NATO's early use of nuclear weapons tactically against any significant conventional attack.[14]

The Conservative Government in the UK saw the credibility of its independent nuclear deterrent threatened by Flexible Response in two ways. First, NATO plans for a protracted conventional war might place Britain in the untenable position of initiating escalation. More immediately, the Labour Party, which went on to win the 1964 general election, was opposed to the independent deterrent and welcomed the McNamara proposals.[15]

The French, at a time of other strains in relations with Washington, were the most critical of McNamara's concepts. Paris viewed the proposal to expand NATO's conventional forces as a concrete manifestation of the uncertainty of the American nuclear guarantee. It therefore validated France's central rationale for development of its independent *force de dissuasion*. Moreover, the idea of preparing for a protracted conventional war ran counter to French strategic thinking. French military planners argued that large land forces and their support infrastructure were attractive targets for Warsaw Pact nuclear attacks. Their own doctrine at the time, much like that of the Soviet Union, emphasized the possibility of immediate use of nuclear weapons in response to any aggression. The French argued that their small nuclear force, which could only be used for counter-value strikes against cities and industrial potential, provided a very stable and credible deterrent to aggression.[16]

An acrimonious debate within official circles over Flexible Response continued for nearly five years, during which time France withdrew from the NATO integrated military structure. Finally, in December 1967, NATO ministers adopted MC 14/3, a fairly ambiguous formulation of a doctrine of 'flexibility of response'. This document mandates three general types of military response to aggression, to be applied consecutively or simultaneously:

1. direct defence, which seeks to prevent an aggressor from reaching his objective at the level of military conflict chosen by him (usually taken to mean the early phases of a conventional conflict but not excluding limited nuclear use);
2. deliberate escalation, which is designed to repulse an attack by changing the quality of defensive operations through the use of nuclear weapons or by expanding the regional scope of the conflict;
3. general nuclear response, which implies a full-scale attack with Allied strategic nuclear weapons against the strategic assets of an aggressor.

The synergism of these three principles is meant to impress the Warsaw Pact that NATO is willing and able to escalate a conflict to a point at which the Soviet Union would be at a distinct disadvantage.[17]

This formulation left much ambiguity about the specific tasks that NATO's conventional forces might have to undertake. Official pronouncements on this question stress flexible response based on an integrated web of conventional, theatre nuclear and strategic forces. However, European members of the Alliance have generally focused on the notion that the very threat of escalation is likely to deter any Soviet aggression in the first place. This deliberate ambiguity was designed to allow West Europeans to view MC 14/3 as mandating a brief conventional 'pause' in a conflict before the use of nuclear weapons. The Allies appeared to agree that direct defence was more than a simple holding action before nuclear retaliation, which had been the principal mission of conventional forces under MC 14/2. However, the West European members of NATO concluded that such a posture would require only a modest strengthening of the 'conventional tripwire' to nuclear escalation. This situation is a far cry from the initial American aspiration that Flexible Response would ultimately result in development of capabilities for an indefinitely protracted conventional defence. But the European Allies, particularly the French, did not want to endorse an open-ended conventional defence because they feared that it would erode the willingness of the US to extend to Europe the protection of its strategic nuclear umbrella. The European Allies' view of the more limited requirements of MC 14/3 has guided their defence planning for the past two decades and is central to any consideration of the types of conventional force improvements they might support in the future.

Conventional Defence Efforts since 1967

Despite enduring problems and tensions, Allied interest in the possibility of political *rapprochement* with the East was growing at this time. The Harmel Report on the future of the Alliance was commissioned by NATO ministers in 1966 to explore how a dialogue with the Warsaw Pact could be pursued in order to lower the cost of maintaining security in Europe and to create a climate of greater mutual trust for the resolution of post-war political problems. The Report, adopted by the NAC a year later, concluded that NATO's security should be founded not just on military preparations, but on the 'twin pillars' of defence and *détente*.[18] The Soviet Union had been making overtures to Western Europe for improved relations throughout the late 1960s, proposing the initiation of an all-European conference on security and ratification of the post-war territorial status quo as means towards this end.

During the same period, domestic pressure for a reduction of the American military presence in Western Europe was mounting as a consequence of economic problems, the escalation of the Vietnam War, and French withdrawal from NATO's integrated military command. Between 1967 and 1972 Senator Mike Mansfield introduced a series of non-binding resolutions in Congress calling for a unilateral withdrawal of some US troops from Europe. Mansfield and an increasing number of his colleagues found the maintenance of 300 000 US troops in Europe nearly a quarter of a century after the end of the Second World War to be an unwarranted drain on American resources. In Mansfield's view this presence was more a reflection of the overextension of American military power than a requirement of the defence of Western Europe. The Johnson and Nixon Administrations were able, with some effort, to defeat the several Mansfield resolutions. In an effort to undercut these unilateral measures, the two Administrations sought Allied support for opening discussions with the Warsaw Pact with a view to mutual reductions of forces in Central Europe. At their June 1968 meeting at Reykjavik and repeated intervals thereafter, the NAC invited the Warsaw Pact to begin such negotiations and agreed among themselves that military forces should not be reduced except 'as part of a pattern of mutual force reductions balanced in scope and timing'.[19]

The Allies achieved impressive growth in conventional military capabilities during the early 1960s. The US embarked on an extensive modernization programme, and the German *Bundeswehr* grew quickly to 12 divisions, bolstering the strength of the Central Region.[20] By the late 1960s, however, the US was beginning to divert manpower and other resources to the war in Vietnam. Given this situation, Washington could not supply the political leadership that a sustained NATO conventional build-up would require. Even the deployment of five additional Soviet divisions along the West German border after Moscow's 1968 invasion of Czechoslovakia did not stimulate any significant increases in force levels.[21] Nevertheless overall European spending on conventional forces (primarily for various modernization programmes) actually increased in real terms by an average of nearly 3 per cent per annum during the 1970s.[22] However, much of the motivation for this came from fears of unilateral American withdrawals as a consequence of Congressional pressure and the aforementioned demands of the Vietnam War. Thus the European members of NATO, anxious to demonstrate a willingness to bear a greater share of the defence burden, put their efforts into a number of political and military initiatives.

In 1969 NATO Ministers initiated a review of defence problems confronting the Alliance in the decade ahead, the 'AD 70' study.[23] The study noted that, while strategic and theatre nuclear forces would remain key elements of NATO's security, improvements to Allied conventional capabilities warranted special attention. The study identified eight priority areas for action: armour and anti-armour; air defence, especially hardened aircraft shelters; maritime surveillance and Anti-Submarine Warfare (ASW); maldeployment in the Central Region; the flanks; mobilization; communications; and war reserve stocks (with a ninth, electronic warfare, added in 1971). In tandem with this effort, the members of the EUROG-ROUP, an informal grouping of all European NATO governments except France and Iceland, initiated a special European Defence Improvement Programme to improve allied capabilities in some of the areas identified in the AD 70 study.

The AD 70 process yieded some noteworthy results, such as the allocation of common infrastructure funding for an airfield shelter programme and the development of the NATO Integrated Communication System. However, it was clear that AD 70 was a political 'lucky dip' and that another instrument was needed to guide concrete conventional defence improvements. Thus the Defence Planning Committee (DPC) decided in 1973 to focus on six basic issues: aircraft shelters; anti-armour weapons; war reserve stocks; electronic warfare; mobile air defence; and modern air-delivered munitions. This was one of the first attempts to identify the principal shortcomings of Allied general purpose forces and to put the collective weight of NATO behind national measures to correct them. The one critical flaw of the AD 70 programme was the lack of a mechanism for monitoring implementation. National defence ministries still decided the extent to which they would proceed with the common programme guidelines.

In response to these military and diplomatic initiatives, President Nixon pledged to maintain and enhance US forces in Europe and to refrain from any reductions in force levels other than as part of reciprocal East–West arrangements. Once the Soviets agreed to begin negotiations on Mutual and Balanced Force Reductions (MBFR) in Central Europe in 1973, any unilateral changes in force structures had to be reviewed in the light of arms control considerations.

Conventional defence modernization efforts in the mid-1970s were tempered by the high inflation and low economic growth that followed the 1973 oil crisis and by high hopes about the dividends of *détente*. With the option of spending more on high-cost conventional forces removed, a number of prominent analysts focused attention on getting more from existing defence resources. As Steven Canby argued, since NATO was spending more on defence than the Warsaw Pact and had equivalent military manpower, its conventional inferiority could only be a consequence of misallocation of resources.[24] Greater standardization and interoperability of NATO equipment, reforms of national procurement policies, and restructuring of NATO forces were all advanced during this period as potential ways to glean more from current outlays. However, these proposals appeared to have little impact on national defence programmes. NATO ministerial communiqués paid lip service to the goal of expanding standardization and interoperability, but few concrete results were achieved.

There was some progress in expanding collaborative weapons procurement during this period. In 1972 the EUROGROUP agreed on guidelines for equipment collaboration and identified eight potential areas of endeavour. These initiatives were stimulated by desires to avoid unnecessary duplication and to redress the imbalance in transatlantic defence trade. They hoped that the cost of European equipment would decline because of improved economies of scale, thereby making them more competitive in the marketplace and more numerous on the battlefield. As many of these projects evolved, it was clear that political and industrial considerations in various countries often undercut hopes of reducing unit and R&D costs.

The Long-Term Defence Programme

By the mid-1970s, the stalemate in the two principal arms control talks, SALT and MBFR, coupled with the alarm about Soviet involvement in Third World conflicts, led to declining expectations in NATO capitals about the potential dividends of *détente*. In addition, several influential studies of NATO's conventional capabilities and readiness focused attention on the steady growth of Soviet military power. The collective judgements of the Hollingsworth Report on the US Army's conventional war-fighting capability in Central

Europe, the Nunn–Bartlett report on NATO's capabilities against the emerging Soviet threat, RAND Corporation studies on conventional defence, and flexibility studies carried out by the Supreme Headquarters, Allied Powers Europe (SHAPE) were reflected in the DPC's December 1976 communiqué, which concluded that NATO needed to take 'further measures ... to reverse effectively the adverse trends in the NATO–Warsaw Pact conventional military balance'.[25]

Restoration of Alliance relations as the focal point of American foreign policy was one of the priorities of the Carter Administration when it entered the White House in 1977. To that end, the new Administration was also committed to enhancing US military capabilities in Europe. The European NATO governments were receptive to these two initiatives and urged that the military programme emphasize non-nuclear forces.

By mid-1977 President Carter had secured the commitment of his European counterparts to begin a long-term defence programme and to strive for real growth of approximately 3 per cent each year in military spending in order to realize it.[26] The idea of a percentage pledge had been circulating in NATO for some months, and 3 per cent real growth was the favoured figure. Most European governments had achieved growth in defence spending of this magnitude through the mid-1970s.[27] This figure was also attractive to Washington because the US GNP was projected to grow at about this rate, which meant that military outlays could increase without allocating a higher proportion of national resources for defence. Finally, the 3 per cent growth idea could be presented as a prudent NATO matching of the levels of growth that the Central Intelligence Agency (CIA) had estimated for the Soviet Union. The Carter Administration also called for set of 'quick fix' improvements in the 'three Rs' – readiness, reinforcement and rationalization – that had been identified in the SHAPE flexibility studies. President Carter emphasized the importance of cooperation on arms procurement and recognized the need for a more balanced 'two-way street' in arms sales with the European partners.[28] This defence plan bargain exchanged European commitment to burden-sharing for an American pledge to achieve a better balance of military trade.

The fully-developed Long-Term Defence Programme (LTDP) was endorsed by the NATO Heads of State in May 1978. The LTDP identified ten functional areas as priorities for improvements. The tenth, concerning theatre nuclear weapons, received the most public

attention. However, the other nine 'action areas' concerned convention
al forces: readiness; reinforcement; reserve mobilization; maritime
posture; air defence; command, control and communications (C^3);
electronic warfare (EW); rationalization; and logistics. The Programme was designed to yield more efficient use of defence resources
by expanding the levels of coordination, joint planning and mutual
support.[29] It was expected that the LTDP could stimulate joint
Alliance initiatives in the development of new families of weapons
and standardized equipment.

The LTDP's nine conventional 'action areas' were very similar to
the AD 70 'key problem areas'. Most of NATO's shortcomings in the
early 1970s had not been rectified. The LTDP did, however, place
new emphasis on rapid reinforcement, combat logistics and the
rationalization of armaments production. A key new feature of the
LTDP was the 'family of weapons' concept in which each NATO
country would take the lead in developing a member of a family of
weapons with co-production of each other's designs. This principle
was meant to allay European fears that standardization and
reduction of duplicate R&D would be accompanied by renewed
pressure to buy American weapons. Since the US would be excluded
from the development of certain weapons, a market for European
'family members' would be assured.

The actual achievements of the LTDP in redressing NATO's
deficiencies and the level of growth in Allied defence spending have
fallen well short of established goals. This outcome is a consequence
of increasing budget pressures after the 1979 recession, doubts about
the effectiveness of new weapons, flaws in the design and implementation of the programme, and diversion of political attention to the
nuclear force modernization question.

The scope of the LTDP was too broad. Not all the NATO
countries could afford to move on the specific measures called for
under each general 'action area'. Too many items were accorded the
highest urgency, making it difficult for defence ministries to decide on
priorities or make trade-offs. The programme required Washington
to make far fewer adjustments in its defence plans than were required
of most of the Allies.[30] The LTDP can also be criticized for not being
sufficiently comprehensive; a number of areas of concern, such as
shipbulding and aircraft, were not addressed by the programme.
However, expanded infrastructure budgets were part of the programme in the late 1970s.

After the December 1979 Soviet invasion of Afghanistan, Allied military commanders marked certain aspects of the LTDP and country force goals for accelerated action. In May 1980 the DPC urged that special priority be given to what became known as the 'Afghanistan Phase I' measures: war reserve ammunition stocks; electronic warfare; and nuclear, biological and chemical (NBC) defence. The Soviet move into South-West Asia underscored the fact that the US might have to use European-based forces in defence of Western interests outside the NATO Treaty area. To offset such redeployments, the US presented 'Afghanistan Phase II' measures to address problems in force readiness, reserve mobilization, war reserve stocks, reinforced airlift, maritime defence, host nation support, and military aid to Turkey and Portugal. The DPC endorsed these measures in December 1980.[31]

Thus the scope and pace of the LTDP became subject to *ad hoc* alterations which disrupted national defence plans and reduced political support for the programme. More importantly, only a few of the NATO countries consistently realized the financial commitments of the programme. Indeed, only five countries achieved real spending increases in the vicinity of 3 per cent during the LTDP's first year. Non-US defence outlays increased in 1980 and 1981 by an average of 2.7 per cent in real terms, but only 2.3 per cent in 1982 and 1.2 per cent in 1984.[32] These spending trends have helped to create the popular image in the US that the European members are not taking their fair share of the defence burden.

The 3 per cent 'solution' caused some disharmony within the Alliance. There is no question that the US was sometimes heavy-handed in its effort to obtain 3 per cent growth in defence budgets and that there has been an excessive focus on this dubious measure of defence effort. Similarly, the US has not lived up to its parallel undertaking that, in tandem with the LTDP, it would do more to enlarge the 'two-way street' in arms procurement. None the less the fact remains that, even under this pressure, the European Allies have been unwilling to increase defence spending in the midst of the prolonged economic recession. Overall, only about 70 per cent of priority defence improvement initiatives identified in NATO's formal 'force goals' have been fulfilled in recent years.

Despite the LTDP's limited programmatic success, it did include the introduction of some useful mechanisms to improve NATO long-range planning. The limitations of the complicated NATO defence planning cycle are well known.[33] The five-year planning cycle is too short to have an impact on national budgets and specific weapons programmes with similar or longer gestation periods. The force goals generally reflect an amalgamation of national plans rather than NATO-wide needs. In this regard, the LTDP's focus on fewer programme areas and objectives projected 10–20 years in the future made it possible to adjust national programmes to reflect NATO goals better. Similarly, the independent programme monitors' annual reviews of implementation in their areas helped to maintain a high level of visibility.

The LTDP, however, did some damage to Alliance solidarity. By raising expectations of sweeping changes in policy and resource allocation that were never fulfilled, the programme gave rise to a certain amount of recrimination among Alliance partners. Nevertheless, the LTDP did focus attention on conventional force issues and paved the way for the consideration of a variety of new initiatives in the conventional sphere over the past few years.

This brief history induces a certain sense of *déjà vu* when considering the more recent efforts to enhance conventional forces. Before turning to a review of these initiatives, however, it will be useful to outline briefly the political considerations and those contemporary economic and demographic trends which define and constrain the resources available for defence improvements.

THE CONTEMPORARY CONTEXT

The Political Climate

While support for NATO and a strong defence remain quite high throughout Europe, fundamental aspects of Alliance policy were 'repoliticized' as a consequence of European discontent with the conduct and substance of the Reagan Administration's nuclear weapons policies and general approach to the East. [34] This situation is certain to give greater salience to defence policy and complicate the generation of support for NATO force modernization efforts in the coming years. As recent public opinion data reveals, profound doubts exist, particularly in Europe, about the utility of increased or more effective military capabilities. A concrete manifestation of this trend is the European public's scepticism about the need for increased defence spending, along with the perception that it is the very accumulation of weapons on both sides rather than the Soviet Union itself that is the greatest threat to peace. Significant majorities on both sides of the Atlantic see East and West as overarmed and increasingly likely to stumble into war as a consequence of an arms race. Thus only programmes that appear to this large audience to reduce the risks of deterrence failure are likely to gain support.[35]

There is also the problem of recurrent discontent in the US with the relative size of the American defence burden, as compared to those of its European Allies. As an expression of this concern, the Fiscal Year (FY) 1983 US Defense Authorization Bill included a Congressionally-imposed ceiling on American force levels in Germany. Since US forces had returned to the level maintained before the Vietnam War, this action had little impact. However, a real jolt was delivered to the Europeans in 1984 by Senator Sam Nunn, generally regarded as an Atlanticist, who came alarmingly close to obtaining Senate approval for a measure that would have forced a reduction of US troops in Europe if the Allies did not achieve certain conventional defence improvements.[36] The Nunn Amendment was only intended to signal a warning of the depth of American discontent with the European level of effort, and the Senator has not reintroduced it. However, it seems likely that Congress will reapply pressure on the European Allies for greater defence efforts in the coming years for several reasons. The impact of the Gramm–Rudman–Hollings deficit reduction legislation requiring defined levels of annual progress towards a balanced US budget continues to take its toll of American military readiness and procurement expenditures. The Reagan Administration's successor will also be looking for areas to trim expenses, and NATO deployments remain an attractive target.

Despite these legislative actions and recent renewed calls by several prominent officials for reducing the level of US troops in Europe, large majorities among the American public and political leadership continue to favour maintaining or increasing the size of the current commitment. Zbigniew Brzezinski's and Henry Kissinger's calls for the withdrawal of upwards of 100 000 US troops from Europe as a means to stimulate greater West European defence efforts have attracted little support. According to a 1986 poll by the Chicago Council on Foreign Relations, 70 per cent of the American public and 85 per cent of the US leadership favour maintaining or increasing the US military presence in Europe.[37] Moreover, among the members of Congress surveyed, 96 per cent favoured the current commitment. Only 16 per cent of the public and 14 per cent of the leadership sample supported partial or complete withdrawal of American forces.

The Economic Situation

Despite the recovery of the mid-1980s, most NATO countries still confront serious economic problems such as high unemployment and inflation. These countries are also burdened with rising interest rates and mounting government debts, which are in large measure a consequence of increased military spending in the 1980s and oil price rises in the late 1970s. Most of the Alliance partners found it difficult to sustain real growth in defence outlays during the 1980–2 recession. None the less, half the NATO member states were able to increase defence accounts between 1979 and 1986.

As a consequence of this situation and the aforementioned political resistance to increased military budgets, projections for the next few years show little or no growth in European defence spending. While NATO ministers agreed in June 1983 to maintain the goal of 3 per cent real growth until 1990, it was a hollow promise. The UK has declined to be committed to the goal any longer. West Germany is tied to a 3.8 per cent net increase over the next few years, which means little, if any, real growth given projected inflation rates. US defence spending is unlikely to continue to grow as much in real terms as it did in the 1980–6 period, with an average annual increase of 9 per cent. Indeed, Congress is reluctant to maintain this level of growth and has enacted deficit reduction legislation which will also restrain military spending. France stands apart from the other leading Western powers with an increase of 7 per cent called for during the first year of the 1987–91 defence programme, with nearly a third of this funding devoted to nuclear force modernization. However, inflation in France has been running above 5 per cent. Economic and political pressures in other NATO countries are likely to preclude any real growth in defence spending.[38]

Dwindling Manpower

Demographic trends in all NATO countries will result in a dwindling manpower pool for military service, most acutely in West Germany and Denmark. As of 1983 the Western countries had a fertility rate of 1.76, which is 15 per cent below the replacement level. West

German and Danish women are at the low end of this scale, bearing a lifetime average of only 1.3 children.[39] This is not an entirely new phenomenon. Most NATO countries have had fertility rates below replacement rates for over a decade. Thus the number of men in the age group best suited for combat roles (17–30 years) is declining rapidly. At the same time, manpower pools are ageing, with more men in the age group better suited for combat support roles (30–44).

Manpower resources suitable for military combat roles are expected to fall by 50 per cent in West Germany by 1999. Without even more controversial measures than have already been undertaken – extending conscription to 18 months and encouraging career volunteers – the *Bundeswehr* could shrink from its present peacetime strength of 495 000 to fewer than 410 000 troops by the end of the decade and to 300 000 by the end of the 1990s.[40] A decrease in the number of brigades in the *Bundeswehr* would reduce NATO's force-to-space ratios in corps sectors that are critical to the maintenance of a strong Forward Defence.

Solutions to these demographic problems are not in prospect. The 1983 French and the subsequently modified 1981 British defence reorganization plans both assumed lower force levels and hope to realize savings from reduced manpower costs, which could be allocated to equipment procurement. This expectation of replacing manpower with firepower, which is at the heart of the French reorganization plan, is probably mistaken. In the first place, the resources are unlikely to be available to procure all the equipment that is programmed in the current law. In addition, some of the new high-technology weapons systems may actually require somewhat greater manpower to operate, mostly in support functions, than existing weapons with less-complicated C^3 requirements.[41] Another step being given increased attention is recruiting more women to fill support elements, but the extent to which this can be a socially acceptable and militarily useful measure in operational circumstances is at least debatable and probably limited.

DOING MORE WITH SCARCER RESOURCES

Given the economic and demographic constraints discussed in this chapter, NATO military planners and many strategists outside official circles have grappled with various schemes for 'doing more with less'. There have been several mainstream, often complementary, approaches to dealing with these constraints. One option, advocated primarily by officials, focuses on the potential savings and enhancements of force effectiveness that could be realized from greater efficiencies in national efforts and from enhanced collaboration among the Allies in many areas of defence planning and armaments development, particularly with respect to the exploitation of so-called emerging technology (ET) weapons systems. Another group of analysts project that NATO could easily mount the robust conventional defence it desires by better rationalization of existing resources and more effective use of reserve mobilization. Two sweeping proposals for conventional defence improvement would require significant changes in NATO doctrine and are therefore treated in the discussion of strategic concepts for conventional defence in Chapter 7. One group of proposals for a non-provocative defence combines elements of territorial defence strategy with new conventional weapons technologies and/or barriers. Another concept, advanced by American strategists, calls for the adoption of a 'retaliatory counter-offensive strategy'. None of these approaches, as elaborated below, has received overwhelming support on either side of the Atlantic.

Thus, whatever options are chosen, it is clear NATO will have to improve its conventional military capabilities in an environment of marginal political support and tightly constrained resources. The next four chapters examine the broad array of proposals that have been advanced to improve NATO's conventional defences. They assess the operational cost-effectiveness and political acceptability of such possibilities as weapons procurement collaboration, exploitation of new technologies, expanded defence cooperation and innovative tactics that are designed to meet this goal. Chapter 7 then offers a guide to consideration of which of these options are best suited to various strategic concepts for conventional deterrence.

3 Opportunities for Cooperation

THE POTENTIAL CONTRIBUTION OF ARMAMENTS COOPERATION

NATO Ministerial Communiqués have repeatedly recognized the need for greater burden-sharing, rationalization, specialization and common procurement programmes. Some genuine progress has been achieved on all these fronts. None the less, the potential contribution of such undertakings has barely been scratched because of conflicting national political and economic considerations. The principal impediments to more effective cooperation remain domestic economic and political pressures and the very limited degree of coordinated defence planning that exists today.

Considerable resources and personnel could be made available for other purposes if redundancies of effort among NATO members in the areas of logistics, training and equipment were minimized. It has been estimated that NATO capabilities are diminished by 30–40 per cent due to non-standardization of equipment and duplicative research and procurement costs.[1] One study calculated that, in and ideal case, two states entering into a collaborative programme could reduce production costs by 10 per cent as a consequence of economies of scale not available to national enterprises, and the pooling of technical expertise could lower each nation's R&D costs by up to 35 per cent although total combined R&D costs would be higher.[2] More effective multinational defence collaboration endeavours would foster a greater degree of commonality and interoperability than exists today. Successful collaborative ventures can also have less tangible but important political benefits: in particular, the deepening of Allied integration.

The disappointing history of multinational weapons collaboration within NATO, however, does not generate optimism about its potential in the future. The formal NATO structures for collaboration, the Standing Conference of National Armaments Directors (CNAD) and the Independent European Programme Group (IEPG), have yielded few concrete results, and most ventures have been carried on outside these frameworks.[3] Of the more than 50 collaborative equipment programmes that have been carried out by various groupings of NATO nations during the past few decades, the most productive have been bilateral or trilateral projects such as the *Jaguar* and *Alphajet* aircraft and the *Milan* and *Roland* missiles. There have been a number of expensive failures, such as the Anglo–German–Italian SP70 self-propelled 155 mm howitzer project, and even most of the successes have fallen short of the ideal. In some cases, R&D costs have been an estimated 30 per cent higher than a national venture would have been. Development timescales have similarly been prolonged. Moreover, even some of the greater technical successes have not delivered expected economic or political benefits and have sometimes even exacerbated tensions within the Alliance. For example, the very impressive European *Tornado* multi-role combat aircraft has seen tremendous cost growth that has limited procurement, and the F-16 fighters assembled by the European consortium have higher unit costs than aircraft produced in the US. Government involvement has often slowed the pace of development, as have squabbles over design and share of the work. Indeed, the number of such ventures slackened off considerably in the later 1970s because of general disillusionment with their utility.

Renewed Interest

During the last few years, however, interest in weapons collaboration has been rekindled for several reasons. The continuing economic stagnation in most of Europe and the seemingly inexorable growth in the cost of new weapons systems have forced several defence ministries to recognize that force modernization in the 1980s will be severely constrained if solely national procurement programmes are pursued. Defence equipment costs, for example, have been rising faster than the rate of inflation in the rest of the economy with annual real increases of the order of 6–10 per cent. Political considerations have provided added impetus to many European ventures.

The US has been championing weapons collaboration not only for many of its aforementioned economic benefits but also as a means to spur the development of more effective European conventional defence capabilities. The Reagan Administration has made armaments cooperation, as part of an overall 'resources strategy', a priority in NATO discussions in the aftermath of the INF deployments.[4] US Under-Secretary of Defense Richard DeLauer called on the Defense Science Board to form a task force on industry-to-industry armaments cooperation. Chaired by Malcolm Currie, one of DeLauer's predecessors, the Task Force Report concluded that industrial cooperation between the US and its NATO partners would strengthen the Alliance at the cost of increased competition for American industry. But it noted that US technological leadership is deteriorating in any event, and that investment for long-range R&D must be increased.

Pressure from the US Congress has been mounting as well. The 1982 Roth–Glenn–Nunn Amendment called on Allied Heads of Government to agree on a strategy and a structure for improving Alliance arms cooperation, as well as policies to end duplication and promote burden-sharing. In addition, the Gramm–Rudman Deficit Reduction Act of 1985 has helped to galvanize Congressional support for cooperation. The defence spending cuts that this budget-balancing legislation will impose, coupled with growing interest in improving NATO's conventional capabilities (especially the European contribution), are likely to ensure that armaments cooperation remains an American priority.

Indeed, these very concerns gave impetus to the Nunn–Roth–Warner Amendments to the Defense Authorization Bills which allocated $125 million in FY 1986 and $185 million in FY 1987 for cooperative development projects among NATO countries. The 1987 five-year US Defense Plan allocates a total of $2.9 billion for cooperative R&D projects with the Allies. By mid-1987, 12 NATO governments had signed letters of intent to develop seven weapons systems that will be partially funded by the Nunn Amendment. These systems include: terminally-guided 155 mm artillery ammunition, modular stand-off weapons containers, an advanced airborne radar system (SOARDS), an identification of friend of foe (IFF) system for combat aircraft; and three computer and information distribution projects. Cooperation on these projects is spread widely among the participants, and harmonization of requirements has proved difficult. These programmes are not particularly glamorous, but they could have a significant impact on certain military capabilities. For example, the new communications system would facilitate more effective use of a variety of weapons.[5]

The 1985 Nunn Amendment also established a mechanism for considering these projects at early stages of the acquisition process, by requiring the Pentagon to set aside funds ($40 million in 1987) to test European equipment against American competitors before moving to production of any new system. Other funds have also been made available in the Foreign Weapons Evaluation programme for Army and Air Force side-by-side testing. Between 1980 and 1985 more than $50 million was spent on this effort, resulting in US procurement of more than $2 billion worth of foreign military equipment. Congress has made it clear that it believes the time has come to develop the two-way street, and with good reason. The side-by-side programme could result in substantial savings in the US defence budget. For example, if the programme had been in effect earlier the US might have procured one of the proven European air defence weapons rather than wasting $1.9 billion on development of the DIVAD system, which was ultimately cancelled because of poor performance and excessive cost.

Washington has been somewhat disappointed with the slow pace of European efforts to pursue these projects. Despite the letters of intent, none of the European governments have allocated funds to cover their share of the costs of these programmes, as is required under the Nunn Amendment. Some NATO officials argue that this delay reflects the problem most European governments have of securing parliamentary and other authority to alter defence spending plans which have been previously negotiated.[6] These projects have achieved a high degree of political salience within NATO and the participating countries. Several collaborative projects with the US, such as the multi-launch rocket system (MLRS) and the F-16 fighter programme, have also been expanded. The course all these initiatives take during the late 1980s will probably determine whether any new projects are launched.

New life has also been injected into the NATO weapons collaboration mechanisms, the CNAD and the IEPG. The CNAD has become the focal point of NATO planning for development of the common frigate, exploitation of emerging technology, and expansion of industry-to-industry cooperation. The IEPG held its first full ministerial meeting in its ten-year history in November 1985, and has held several since then. The IEPG is moving to seize some long-term collaborative opportunities. It has identified 30 programmes for possible action and has made some progress on five of these. French hopes of transforming the IEPG into a focal point of collaborative weapons programmes may yet be realized. The NATO Independent Defence Study Team recommended in early 1986 that the European members of NATO develop a common armaments market and purchase only European-manufactured equipment, even if less expensive items can be secured elsewhere.[7] The Study Team argued that this was one way to realize an aggregate market size that would make collaborative efforts more feasible and cost-effective. Whatever organizational vehicle is chosen, the fate of joint weapons projects will be determined by the political will among Allied governments to deal with the considerable financial and industrial interests that have so often frustrated progress in this area. In times of slow economic growth, domestic pressures to support national industries first will surface and, unless firmly resisted, are likely to undermine progress on this front.

Expanded European Efforts

There has been a resurgence of cooperative projects involving two or three European countries. West European governments and industries have pursued five major collaborative projects during the 1980s, with mixed results. These have included joint efforts to develop an air superiority fighter, two helicopter programmes, an airborne warning and control aircraft, a refuelling aircraft, and medium and light transport aircraft. It is, however, difficult to be very optimistic about the possible dividends of these ventures.

In 1983, France and West Germany signed an agreement to develop and produce the PAH-2 attack/support helicopter. The PAH-2 project has experienced some difficult periods largely because the two nations have had major problems in harmonizing their requirements sufficiently to reduce development costs. The design problems were resolved in March 1987. Instead of the three originally envisaged, there will now be a single anti-tank model, with French and West German variations, and a support (anti-helicopter) model which will be used by the French *Force d'Action Rapide*. The *Bundeswehr* will purchase 212 of the anti-tank variant for an expected total cost of $425 million and the French will buy 140 anti-tank and 75 support aircraft for more than $525 million. These systems were to have been armed with a third-generation anti-tank missile to be developed jointly by a British–German–French consortium. However, the German anti-tank helicopter will now be equipped with *HOT-2* and the French version will carry the AC-3G.[8]

Several European governments (France, West Germany, Britain, Italy and Spain) encountered many of the familiar impediments to cooperation in the development of the future European fighter aircraft (EFA). While there were some genuine differences over military requirements, the EFA programme came apart, as did other projects before it, because of division over export policy. The French entry in the competition, Dassault's ACX, was considerably less sophisticated and less expensive than other contenders, a design decision that appears to have been made to strengthen the plane's potential export market outside Europe.[9] Arms exports are an increasingly important element of the French economy, with some estimates indicating that arms sales comprise 30 per cent of all French industrial exports.[10] The French position is in such striking contrast to the West German policy of restraint in arms sales to the Third World that the EFA project was bound to suffer from differences over the export question. More important, however, was the question of how the design and production work was to be distributed among the participants. French demands on design specifications and production arrangements (France sought 46 per cent of the work) finally forced four of the five partners to proceed without France in a separate EFA programme.[11] After seeking ways to rejoin the group at a much lower share, the French government decided to pursue its own fighter, the *Rafale*.

The French attitude on bilateral and multilateral weapons cooperation was best articulated (perhaps inadvertently) by former Defence Minister Charles Hernu when he said that, while these projects are important if Europe is to cope with intense international competition, they need 'to be done on the condition that each of the participants protect their essential interests.'[12] This opinion is shared by most of his colleagues. Thus, rather than jointly procuring 1000 standardized aircraft at a cost of about $15 million, these countries will probably end up with a mixed force of European and American planes with a much higher price tag.

Another complication is the fact that the European weapons producers are highly dependent on US firms for 'state-of-the-art' electronics and other high-technology components. Thus if the US hopes to foster and participate in various cooperative projects, it will have to provide Europeans with access to critical technologies and to offset the effects of this dependency by purchasing much more European equipment than it has in the past. This requirement is at odds with the Reagan Administration's move to restrict the flow of technology to allies and adversaries alike. It is far from clear how much US technology that has been shared with its European allies in various cooperative armaments and equipment programmes has found its way to the Eastern Bloc.[13] None the less, the Reagan policy treats this as a particularly grave area of leakage and has imposed controls on a number of military and 'dual use' technologies. Secretary Weinberger's 17 January 1984 Directive 2040.2 now restricts the transfer of technology that is categorized as 'sensitive', as well as that which is 'militarily critical'.[14]

These export control policies could slow the development of transatlantic defence cooperation and the 'two-way street' in arms trade. The EEC registered a formal complaint with the US Government in 1984 over proposed amendments to the Export Administration Act which would provide the President with authority to terminate imports from countries that failed to comply with these restrictions. Such punitive measures are seen as an infringement of national sovereignty. The potential for conflict among the Allies in this area is substantial, given the fact that the European members of NATO do not share the Reagan Administration's view of the magnitude of the technology transfer problems.

The prospects for a major expansion of the American market for European defence equipment also appear remote. The balance in transatlantic defence trade has improved over the past few years.[15] The military trade balance between the US and Western Europe still tilts about 4:1 in favour of the US, although the imbalance was 7:1 in 1980. The US Specialty Metals clause severely limits access of foreign arms manufacturers to the American market. The Reagan Administration's industry-to-industry approach to this problem is anathema to the Europeans. Free market competition would badly damage European defence industries, which constitute important sectors of their economies and provide significant employment. At the same time, the US government and industrial firms have objected to European protectionism and export subsidies in key sectors such as aircraft. Thus any deterioration of these and other aspects of transatlantic economic relations could hamper armaments cooperation in the coming decade.

There have been both significant successes, such as the US Army's award of a $4.3 billion contract for its Mobile Subscriber Equipment battlefield communications system (RITA) to the Franco–American team of GTE/Thomson–CSF, and damaging failures, such as the collapse of the *Roland* anti-aircraft missile deal, in building the 'two-way street'.[16] Individual exceptions to the primarily one-way flow emerge when one looks at country breakdowns. In 1984, for example, the US purchased more equipment from Belgium, Norway and France than it sold to these countries, but the transatlantic defence trade balance will never approach 1:1 so long as the US spends more than twice as much as its Allies on military R&D. The European members of NATO will either have to agree to live with this imbalance or spend far more than they presently do on defence research. Defence cooperation has therefore not yielded the dividends hoped for over the past two decades. Despite the renewed interest in multilateral weapons projects, and activity in various NATO planning mechanisms, many of the fundamental political and economic hurdles that have hampered past efforts of this nature remain. Despite the general protectionist sentiment in Congress concerning other trade issues, the US side-by-side testing programme has solid political backing and may improve European penetration of the American market. In contrast, tougher US restrictions on technology transfer are likely to impede the development of both European and transatlantic projects. There may be some modest dividends from armaments cooperation, but NATO will hardly recover the 30–40 per cent of resources that some feel is wasted on duplicative efforts.

The Challenge of Technology

Europe has responded to the new American technological challenge with vigour. The French Government has shown considerable initiative in this area, making its collaborative civilian technology development programme, *Eureka,* and other bilateral and multilateral projects the spearhead of its efforts to develop a greater 'European defence identity' within NATO.[17] The initial proposal of 18 April 1985 was so vague that it appeared to be 'little more than an idea in search of a framework'.[18] Some observers saw France as simply trying to establish itself in a leadership role in the formulation of a European defence identity. Others dismissed *Eureka* as a French effort to minimize European participation in the US SDI programme. However, by the time of the group's December 1986 meeting in Stockholm, 19 West European countries had adopted a governing charter and endorsed a total of 111 research projects with a total value in excess of 4.4 billion European Currency Units (ECU). The charter and several ministerial meetings have clarified the aims of *Eureka,* and member states have begun to resolve many of the fundamental organizational problems that have generally hampered such cooperative efforts. Nevertheless, uncertainties about the durability of the member states' financial commitments and about the feasibility of reconciling quite distinctive industrial and research policies remain.

The primary goal of *Eureka* is to coordinate European research in several broad areas: robotics; information-processing; telecommunications; new materials; biotechnology; lasers; environmental protection; and transportation.[19] The project is supposed to evolve into a coordinating agency to promote joint research programmes and to develop infrastructure and standards. Many European leaders hope that this coordination of their R&D programmes at several levels will enhance competitiveness. Thus, although much of the original impetus for *Eureka* came in response to SDI, it has now become part of a general European interest in keeping up with the US in high technology. *Eureka* and the non-military programmes formulated under the aegis of the European Economic Community (EEC) – for nuclear energy, Euratom; space, ESA; computers, ESPRIT; manufacturing technology, BRITE; nuclear fusion, JET; and telecommunications, RACE – may ultimately result in a more efficient use of European research funds for high technology. However, all these activities will be tempered by strong and often divergent national interests.

Conventional Defence Improvement (CDI) and the Conceptual Military Framework (CMF)

A less ambitious – but still desirable goal – of increased cooperation would be the development of what a 1985 CNAD report noted the Alliance sorely lacked, namely a genuine armaments planning system. The well-massaged force goals approved by NATO defence ministers tend to reflect little more than an amalgamation of independently derived national plans. If the momentum of agreed initiatives is to be maintained, it is vital that they do not become submerged within the hundreds of improvement actions in the often-ignored force goals. Moreover, the long-term character of some of the conventional force improvements that have been proposed, such as complex weapon systems with ten-year development cycles, demands a planning mechanism of comparable scope. This problem may be endemic to the nature of the Alliance. However, better long-range planning mechanisms within NATO, offering guidance that could be more easily factored into national defence planning than can the short-term 'Ministerial Guidance' document, would foster greater military effectiveness and cohesion.

At their December 1984 meeting NATO defence ministers, reflecting the emerging consensus in several member states that the Alliance should redouble its efforts to improve conventional forces, launched a series of measures which have become known as the CDI programme. The Ministers undertook several measures to improve Allied readiness and sustainability. They also authorized the Secretary-General to finalize a CMF, a conventional defence resource allocation strategy that would focus national efforts, and mechanisms to encourage international cooperation in defence procurement.[20]

During early 1985, NATO concluded an extensive review of the state of conventional deterrence. Allied military authorities were charged with development of the CMF which defines the critical warfighting missions of Alliance strategy. The CMF was developed in response to the confusion many Europeans felt with the many CDI proposals emanating from the US during the early 1980s. As West German Defence Minister Woerner suggested on several occasions, there was a need for a conceptual framework to clarify how these new initiatives related to established practices in enhancing NATO's conventional posture. Thus the CMF identifies the central military missions of the Alliance as defeat of the lead echelon of attacking forces, attack of enemy follow-on forces, maintenance of control of the sea lanes and airspace, projection of maritime power, and protection of shipping and rear areas.[21]

At their spring 1985 meeting, NATO defence ministers endorsed a CDI plan comprised of the CMF, a better assessment of the threat, and a determination of critical military deficiencies. The CMF has become a long-term guide in the development of tactics and technology by expanding the Allied military planning horizon from six to twenty years. NATO military commanders now evaluate national forces each year in the light of the CMF with the object of identifying deficient capabilities for priority action. At their meeting early in 1985, NATO defence ministers agreed that the Alliance's most pressing problems were: shortfalls in active forces; inadequate mobilizable reserves; inadequate offensive counter-air capabilities; insufficient ability to attack follow-on forces; the lack of an IFF system for aircraft; inadequate anti-submarine warfare, anti-maritime air, and mine counter-measure capabilities; inadequate ability to support early air reinforcement; shortcomings in the forces of Greece, Portugal, and Turkey; and inadequate stocks of ammunition, fuel and spare parts.[22]

Since the CDI was initiated, NATO has been moving on several fronts to improve conventional forces. At their December 1985 meeting, NATO ministers instructed the CNAD to implement a new Armaments Cooperation Improvement Strategy.[23] Through this plan, the CNAD is trying to coordinate national and multilateral equipment programmes in ways that redress NATO's conventional deficiencies on the basis of established priorities.

It remains to be seen whether the CDI and CMF will have any real influence over national decision-making, unlike a number of earlier planning initiatives which were soon forgotten. The documents' impact may depend on how the Major NATO Commanders (MNC) use them in the development of their more detailed analyses of missions, force goals and technological opportunities. NATO has been using the normal planning process, the force goals, to identify the most urgent shortcomings in each member's forces that should be corrected. Thus far achievement of CDI-highlighted force goals has been uneven.[24]

Concrete results will have to be achieved in the near term if the political momentum behind these initiatives is to be sustained. CNAD met in February 1986 to review the projects to be initiated under this rubric. The CNAD will be reporting to the NAC, and this organizational link will give a high profile to the recommendations. It might be desirable to promote collaboration on some of the other emerging technology items in the CDI that look most promising and seem to enjoy political support.

Coincident with this renaissance of cooperative weapons development and procurement, there has been a revival of some old concepts for shoring up the European pillar of NATO's deterrent and some dramatic new initiatives in the area of military planning and operations. These aspects of NATO's cooperative options will now be examined.

EUROPEAN DEFENCE COOPERATION

Greater cooperation among the European members of NATO in military planning, training and operations could enhance the combat effectiveness and deterrent potential of Allied general purpose forces. However, after the collapse of the EDC idea and the shift to reliance on nuclear deterrence, there was little political momentum to sustain such cooperation. This situation began to change in the early 1980s as a consequence of disillusionment with American security policies, mounting doubts about the viability of NATO's current strategy, and a new search for ways to improve the conventional component of deterrence. With progress on European integration stymied by wrangles over EEC agricultural policy and budgetary contributions, several European leaders began to suggest that cooperation on security matters might provide broader avenues for progress.

This movement was given added impetus by superpower discussions during and after the Reykjavik Summit on deep reductions in nuclear weapons in Europe and by mounting concerns in Allied capitals about the implications of what are seen by many as inevitable American troop withdrawals by the end of the decade. The Reykjavik discussions of nuclear abolition caught many West European governments by surprise. Thus proponents of European defence cooperation have found the course of events bolstering their contention that Europeans will have to shoulder more of the burden of their own defence and to speak with a unified voice in Alliance deliberations if they wish to have greater influence over the evolution of East–West security relations. British Foreign Secretary, Sir Geoffrey Howe, articulated this post-Reykjavik mood most eloquently when he said:

A Europe which gets its ideas straight is a far more rewarding partner for the United States and far more likely to have its views taken seriously than a Europe which speaks with a multitude of voices. If we want our particular European concerns to be clearly perceived and taken into account in negotiations between the United States and the Soviet Union, then we must argue them out clearly among ourselves and come, whenever possible, to a common view.[25]

Since 1983 a number of European and American officials and analysts have been advocating the development of a European 'defence identity' and a strengthening of the solidarity of the second, European pillar of the Alliance.[26] Several European governments have undertaken vigorous reviews of options for expanding defence cooperation, and several concrete initiatives have been launched. This political climate has been conducive to closer military cooperation between Paris and Bonn and may ultimately stimulate a somewhat less ambiguous French role in Allied defence planning, although France is most unlikely to relinquish the option of deciding when and how to engage its conventional and nuclear forces in the defence of Europe.

The Western European Union (WEU)

Calls for a revival of the Western European Union (WEU) as a mechanism to shore up the European pillar of the Alliance have been voiced intermittently over the past two decades. The WEU was established in October 1954, following the failure of the EDC. The WEU's principal function was to provide a framework for overseeing West German rearmament. The WEU Treaty reaffirms the collective security obligations of the seven members (Belgium, France, West Germany, Italy, Luxembourg, the Netherlands and the United Kingdom) to one another under other accords, and requires close cooperation with NATO on military planning. The Treaty also placed certain limits, since lifted, on the sizes of the parties' armed forces. The WEU has two standing bodies, the Agency for Control of Armaments (ACA), which monitored German rearmament, and the Standing Armaments Committee (SAC) which was supposed to promote armaments cooperation and standardization. Other organs include: a ministerial-level council, which until recently met annually; a Permanent Council, chaired by the Secretary-General, which monitors Treaty implementation; and an Assembly of parliamentarians.

During the past 20 years the WEU has been politically moribund, but the French government maintained an intermittent interest in promoting the Paris-based organization as a focal point for European defence cooperation. The possibility of revitalizing the organization was discussed in 1973 and 1975, and eventually, in December 1980, the WEU Assembly adopted a resolution calling for the institution's revival.[27] The most recent efforts to inject life into the moribund organization were initiated by the French Government in late 1983. The French were searching for a mechanism that would allow them to broaden their defence cooperation with West Germany and to play a leadership role in the development of Western security policy. The present push has resulted in the institutionalization of several high level meetings each year and the commissioning of two studies to be prepared by the WEU staff. It appears that the WEU will at least become a more active forum for member governments to exchange views on security policy and to develop a distinctive, more effective European input into NATO policy deliberations.

At its June 1984 meeting in Paris, the WEU's seven foreign ministers decided to take concrete steps to revive the organization.

They agreed to meet again – this time with defence ministers – in Rome to issue a major political document.[28] Ministers also agreed to commission the staffs of their Standing Committee to prepare two studies, one on arms control problems and another on the nature of the Soviet military threat confronting Western Europe. The rationale behind these studies was that Europeans needed independent assessments of various problems in order to develop a more coherent European point of view.

The 'Statement of Rome' that emerged from the defence and foreign ministers' thirtieth anniversary session in October 1984 spoke of a new European solidarity. However, the most significant outcome was the ministers' decision to meet in future twice a year.[29] In addition, the Rome Statement emphasized the WEU's role in developing a consensus among the members on security matters and encouraging further progress towards European integration. The ministers made it clear that defence matters should be left to NATO, but that WEU could help to stimulate increased armaments cooperation by 'encouraging' the activities of the IEPG.[30] The Rome document also noted the goal of developing clear European positions on events outside the North Atlantic Treaty Area. Yet no expansion of the WEU's staff or budget was authorized by the ministers, so it will clearly fall to the Assembly and member governments to energize the body's activities.

The practical problems of forging a European consensus on security issues were most apparent at the first WEU ministerial meeting after the Rome session, which was held in Bonn in April 1985.[31] At the low-profile Bonn meeting, the WEU ministers attempted with little success to agree on a joint approach to the US SDI. There was considerable American pressure to avoid reaching a joint position at that time because it was assumed that it would not be supportive. Nevertheless, efforts to revive the WEU continued and, two and a half years later, members were able to adopt a 'Platform on European Security Interests', a basic statement of common European security goals.

The advantage of the WEU as a forum for European security consultations is that it is 'at once 2 + 5 and 10 − 3'.[32] That is to say, it incorporates the important Franco–German relationship into a larger European context, and its deliberations are not subject to the inhibitions that the Greek, Irish and Danish presence in the political cooperation machinery of the EC impose on that body's consideration of security matters. The British, Dutch and West German

governments, which had retained some reservations about the
WEU's role, became enthusiastic supporters after the Rome meeting.
Italy finds the WEU an attractive forum for developing
European defence cooperation and favours it over the military
relationship between West Germany and France as the motor of
European defence cooperation. Portugal has applied for membership,
and Spain is also interested in joining. Some European officials
still argue that there is little need for another European security
consultative body given the existence of the EUROGROUP, other
special sub-groups in NATO, and the IEPG. Clearly the IEPG is the
more appropriate instrument for co-ordinating defence procurement
efforts. However, over the next decade the WEU may prove a useful
forum for developing European solidarity on arms control and
general security policy, and give some additional impetus to
armaments cooperation.

Other Cooperative Possibilities

In addition to the revival of the WEU there has been an upsurge in
bilateral and multilateral defence cooperation among the West
European members of NATO, and a number of sweeping proposals
have been advanced. The French and German governments have
been expanding their security cooperation, particularly in the area of
conventional operations. The French and British governments have
agreed to closer bilateral consultations on nuclear and conventional
defence and some officials have discussed the possible extension of
the French nuclear umbrella over West Germany. The prospect of
American troop withdrawals from Europe has increased the intensity
of these discussions.

In a 15 January 1987 speech in London on a wide range of
European issues, French President Mitterrand suggested that France
and the United Kingdom attempt to work out joint procurement
efforts and military plans with respect to both nuclear and conventional
weapons, in much the same way as France has been coordinating
conventional plans with West Germany.[33] Two months later French
Defence Minister André Giraud and his British counterpart agreed to
coordinate their countries' military procurement policies and to
consult on the strategic environment.[34] President Mitterrand also
argued in his address that European defence ministers should discuss
the development of a common approach to dealing with 'strictly

European eventualities', which might not trigger 'the whole Alliance machinery'.

There have been equally novel ideas advanced by a number of different officials. In April 1984, a CDU/CSU spokesman in the West German *Bundestag*, Jurgen Todenhoffer, called for the creation of an integrated European nuclear force, including US, French and British weapons, which would be managed by an executive group consisting of the US and all European members of the Alliance.[35] This proposal received a favourable reaction from several West German officials and commentators. As the next chapter elaborates, Franco–German military cooperation has expanded dramatically, and there has been serious discussion in both countries as to whether France might some day be able to offer a nuclear guarantee to its European partner. However, the chances that a joint Franco–British nuclear deterrent would ever be placed under the control of a supranational West European organization seem remote. Nevertheless, unless some such mechanism to share decision-making authority over weapons development and use were achieved, public opinion data suggest that other Europeans would not find a national nuclear guarantee by one or both of the present West European nuclear states very credible.[36]

There has also been candid talk of how Europe might manage its defence in the face of substantial withdrawals of American troops. Such an event might force France to participate in some form of integrated European military command. With redeployment of its 296 000 troops France might be able to substitute for some US forces. Others have suggested that better use of reserve manpower, reinstituting conscription in Britain and lengthening the term of military service in several Continental countries could help offset American withdrawals. But all these proposals represent only partial substitution for active duty American troops.

These calls for united European defence planning and preparations for military activities independent of the US illustrate the growing European determination to achieve increased self-reliance in the defence sphere. This mood seems likely to endure even if Washington's security policy becomes more congenial to Europeans after the Reagan Administration. A concrete illustration of this disposition can be seen in the French discussions with the West Germans on the development of an imaging satellite, based on the French SPOT earth resources mapping system, which would provide intelligence data independent of US reconnaissance systems.[37]

Indeed, as Sir Geoffrey Howe's speech suggested, many Europeans accept Henry Kissinger's argument that the dependency relationship inherent in Europe's current overreliance on extension of the American nuclear deterrent is not conducive to a healthy transatlantic relationship in the long-term.[38] Europe needs to become a fuller partner with the US in charting the course of Western security policy. Before this can happen, Europe must develop its defence identity through more effective coordination of its political insights and substantial military might. Given the enduring geostrategic, economic, and cultural links between Europe and the US, development of a strong European defence identity can only serve to enhance the viability of the Alliance.

4 France, Spain and Conventional Defence

Among the most significant developments in the European security situation during the past few years are the shifts in French military organization and plans and the expanding Franco–German defence dialogue. The Mitterrand government has reorganized the *Armée de Terre*, improving somewhat its capabilities for involvement in the 'forward battle', and has taken a number of other steps that could facilitate a more coordinated French role in NATO military plans. The growing military relationship between Paris and Bonn has become a major factor and a potential driving force for European defence cooperation. Germany and the other NATO Allies are well aware they are unlikely to obtain an unequivocal assurance of active French participation in the forward battle, extension of its nuclear umbrella, or of Allied access to the critical logistical infrastructure on French territory. However, France has taken a number of steps to improve its capabilities to come to the aid of its NATO partners quickly if it so chooses. Despite the undernourished state of its conventional forces (as a consequence of the primacy of nuclear weapons in French strategy), the intervention of these forces, coupled with access to logistical assets on its territory, could improve NATO's conventional defences considerably. Indeed, as Robert Komer has argued, 'whether NATO can achieve a credible nonnuclear initial defense posture in the crucial Center Region may well depend on the key role played by France'.[1]

If NATO military planners could have assured and better-prepared access to the extensive French military infrastructure, and more effective use of Spanish airbases, it would enhance Allied capabilities to sustain a conventional defence. Spain's reluctance to join NATO's integrated military commands makes it unlikely that the country's armed forces will play any significant role in the defence of the Central Region, despite the yearnings of most younger officers for greater involvement in Western defences. Even in the unlikely event that political impediments to Madrid's military integration eroded over time, the Spanish Army and Air Force would require extensive and costly restructuring, training and equipment modernization programmes in order to shift significantly from their current

47

territorial defence roles. However, as its military modernization programme proceeds, Spain may be able to assume some of the tasks presently performed by US naval and air forces based in the country, thereby freeing some American units for other activities.

THE EVOLUTION OF FRENCH DOCTRINE AND PLANS

French national strategy is founded on the notion that France's strategic nuclear forces grant the country a unique and separate security situation in Europe. In this view, nuclear weapons provide France with a deterrent to all potential threats and the option of non-belligerency, often referred to as 'sanctuarization', in any conflict in Europe. Conventional and short-range nuclear forces are therefore viewed as secondary safeguards to give the French president more flexibility in responding to aggression when a strategic response might be unwarranted or to avoid premature use of strategic forces.

Since the early 1960s the concept of the 'two battles' has guided French military planning. In the 'forward battle', to be waged in West Germany, French conventional forces might play a reserve role in support of its NATO Allies. However, the 'decisive battle' for France is the point at which it would make clear its resolve to employ strategic weapons, perhaps by launching a 'pre-strategic' warning shot with tactical nuclear weapons. French conventional forces – primarily the First Army and tactical air forces (FATAC) – could be expected to act in close coordination with the NATO Allies in the forward battle because this action would generally complement the national deterrent.[2] None the less, autonomous actions are still a possibility. France might opt for limited or no participation in the forward battle if it determined that the enemy's objectives were of no concern to it, or it might choose to rely on the stategic deterrent alone to maintain the national sanctuary. On the other hand, France has reserved the right to employ its tactical nuclear forces whenever it becomes necessary to support the 'national deterrent manoeuvre', even during participation in the forward battle. French participation in the forward battle would always be subordinated to the demands of exercising the national deterrent manoeuvre. Consequently French conventional and tactical nuclear forces would remain close to the northern and eastern borders of the country, but could attack the enemy well forward, if necessary. French planning does not envisage a protracted conventional phase of conflict and rejects the flexible response concept as an ineffective deterrent that demands excessive losses of conventional forces.

Thus the purpose of the French 'battle' corps – that is, the First Army and its tactical nuclear weapons supported by nuclear-armed tactical aircraft – is designed to give greater credibility to the national deterrent manoeuvre by avoiding the necessity of early escalation to strategic release. Even if France opted out of the forward battle it might launch conventional and tactical nuclear strikes beyond its borders to warn an aggressor.

The March 1966 memorandum that announced French withdrawal from Allied Command Europe (ACE) suggested that arrangements should be made for liaison between the French command and NATO commands and to determine the circumstances for the participation of French forces in war, should the collective defence provisions of the North Atlantic Treaty be invoked. Paris refused to delineate the circumstances of French involvement. Guidelines for French cooperation with NATO are set out in the 1967 Ailleret–Lemnitzer Agreements, which call for French participation in NATO exercises only in accordance with mutually acceptable contingency plans.[3] Thus French involvement in NATO defences in times of conflict cannot be automatic. Indeed, as Kenneth Hunt aptly noted, French forces would from then on be 'a bonus that might be available'.[4]

There is no question that French withdrawal from ACE and the consequent reduction of troops available to SACEUR weakened NATO's conventional posture in Europe. It may be true that this independent defence posture has enabled the French government to garner public support for a greater defence effort than might have been the case if it had relied more on American guarantees.[5] Nevertheless this posture forced France to devote a larger proportion of its defence resources to nuclear weapons than would have been warranted under the American guarantee to NATO and has reduced the assured level of conventional force capability available to the Alliance.

The other important blow to NATO's conventional capabilities attendant on France's withdrawal from the integrated military command was the removal of nearly all Allied installations from French territory. In addition, access to French airspace and logistical infrastructure became uncertain. This loss of assured access forced NATO to restructure the lines of communication (LOC) in the Central Region. This situation, coupled with Austrian and Swiss neutrality, divided NATO into two halves, and rendered communication and transportation between NATO Allies north and south of this zone much more difficult.[6] Similarly, loss of access to French airspace and airfields, which are generally no longer capable of servicing Allied aircraft, further constrains NATO dispersal of aircraft in crises. Loss of access to French territory might not be critical in a short intense conflict, but it is far from evident that a war in Europe would be of that type. This loss would have more serious consequences for a longer conflict since without France NATO has no rear zone.

The details of the extent of French cooperation with NATO remain classified and somewhat murky. A number of Allied officials, including former SACEUR, General Rogers, have repeatedly stated that French cooperation with NATO over the past several years has been excellent.[7] Nevertheless, France's role remains encumbered by her insistence on no automatic commitment of forces, no peacetime responsibility for ground, sea or air zones, no participation in the forward battle, and maintenance of French forces under national command even if France decided to join NATO in combat.[8]

Maintenance of the concept of a national sanctuary and the independent national deterrent manoeuvre have remained fundamental components of French security policy. The 1972 White Paper on defence included some vague references to the fact that Western Europe as a whole benefits indirectly from French strategy and that France's vital interests include the surrounding areas.[9] However, the French policy consensus emphasized independence and eschewed any firm commitments to European defence. In the mid-1970s President Giscard d'Estaing and the Chief of Staff of the Armed Forces, General Méry, attempted to alter French doctrine by recognizing the increasing importance of forward defence. Méry argued that France could never be 'an absolute sanctuary' isolated from any conflict in Europe by the protection of her strategic weapons. He advocated a posture of 'enlarged sanctuarization' by developing the conventional capabilities to intervene with all or part of the Armed Forces throughout the region from which French security could be threatened, namely Europe and the Mediterranean.

Méry also suggested that, while France could not man the forward battle area in peacetime, it could not stand aloof from this struggle in which French security would be at stake. He argued that France's tactical nuclear weapons could be used to change the course of the forward battle as well as in the pre-strategic warning mode, and that second-echelon participation in the forward battle would enhance

French security while maintaining the option of retreat to her frontiers. [10] This effort to nudge France towards a closer commitment to her allies met with stiff domestic political opposition and forced the Government to deny any change and to adopt ambiguous public pronouncements.

Despite the continuing dominance of nuclear forces in French strategy, conventional capabilities began to receive more attention under President Giscard d'Estaing's 1977–82 *lois de programmation*. Giscard conducted a military reorganization between 1976 and 1980 in which the five mechanized divisions of the First Army were transformed into 15 smaller, more mobile divisions and a greatly simplified command structure was established, with various independent brigades and regiments. Total Army manpower dropped from 331 500 to 311 200, and French troops in Germany were reduced from 60 000 to 51 200. [11] Giscard also embarked on a major equipment modernization programme that improved the Army's mobility and firepower by the early 1980s. This modernization, coupled with the streamlined organization, compensated for reduced manpower. However, after the reorganization was completed in 1980, the First Army (which comprises the bulk of French forces available for combat in West Germany) included seven armoured divisions, each of only 7000 troops and far fewer tanks than its NATO or Warsaw Pact counterparts. Only two of the seven other (multipurpose infantry) divisions, which comprised the balance of the active Army, were viewed as potential reinforcements for the First Army. [12]

The general trends in Army organization, equipment and doctrine initiated by Giscard d'Estaing continued under President François Mitterrand. Consistent with French doctrine, the 1984–5 and 1985–6 *lois de programmation* gave priority to improving nuclear capabilties, largely at the expense of ground forces. Another Army reorganization was begun by President Mitterrand in 1983. The Socialist Government proposed to maintain the effectiveness of the *Armée de Terre* through various organizational reforms, transferral of its responsibilities for territorial defence to an enlarged *gendarmerie*, and improvement of its mobility and firepower by incorporation of advanced technology. [13] Over the 1984–8 period, the Army has eliminated two armoured divisions and its ranks have shrunk by another 22 000 troops. The Air Force, Navy and other services are also undergoing 5 per cent reductions in manpower.

The critical element of this reorganization is the formation, under a single command, of a 47 000 man *Force d'Action Rapide (FAR)*. This corps-sized unit is comprised of two new and three existing divisions.[14] Based in Eastern France and equipped with about 210 transport/support helicopters and a variety of anti-tank weapons, the *FAR* can be available, at a time and place chosen by the French President, for support of NATO or overseas operations. The 4th Airmobile Division, also called the *Force d'Hélicoptères Anti-Char (FHAC)*, is presented as being the most effective component of the *FAR* for dealing with Warsaw Pact armoured divisions.[15] The first major exercise of the *FAR* in a Central European scenario, '*Fartel*', was held in October 1985. When the 4th Airmobile Division deployed over 300 km using its own helicopters on the first day, communications functioned well as did command and control and cooperation with air support. However, this exercise also revealed the limits of the French Air Force in providing requisite logistical support to the *FAR*, particularly the *FHAC*.[16] Two more exercises took place in 1986 and 1987. The *FAR* is neither an independent army corps nor an autonomous rapid deployment force. Rather, it is a pool of formations under a new combined operations command which, by virtue of their mobility and flexibility, can be used to demonstrate French resolve in the early stages of a European war or bolster other forces engaged in combat.[17]

Despite these organizational and training initiatives, French conventional forces are likely to have somewhat less combat capability in the coming years, and the replacement of outdated equipment will be slowed by budgetary constraints as it was during the 1976–80 programme.[18] While the major elements of Giscard d'Estaing's modernization programme were realized, some procurement was stretched out beyond the programme dates and equipment goals for some units have yet to be met. Nevertheless, valuable gains were made in selected areas over the past decade. The Army has begun to upgrade part of its outdated inventory of tanks and other new armoured vehicles, and to improve its mobility, artillery and tactical C^3 system.[19]

Important quantitative and qualitative shortcomings in French air defences and tactical aviation persist. France has finally decided to procure three to five of the E-3A *Sentry* Airborne Warning and Control System (AWACS) airborne low-altitude detection aircraft, the first of which is scheduled for delivery by 1991. The French Navy has favoured purchase of the E-2C *Hawkeye*, which is better suited for naval missions, while the Air Force and defence ministry were interested in looking at the British *Nimrod* Airborne Early Warning (AEW) aircraft. After the British decided to abandon the *Nimrod* AEW project in late 1986, the two countries agreed to purchase common AWACS forces and share training and maintenance expenses. While its attack characterization capabilities will be improving, the Air Force still lacks sufficient numbers of aircraft to perform assigned air superiority, ground support and air defence missions.[20]

Tight budgets will continue to constrain the effectiveness of French conventional forces in the near term. Former Armed Forces Chief of Staff, Jeannou Lacaze, revealed that his men had to cut their operations by 5 per cent in 1985 for fiscal reasons. With a fifth of the defence budget going to the *Force de Frappe* (especially the submarines), resources are likely to be inadequate for other missions, despite the Chirac Government's commitment to improving conventional defences.[21] The 1987 French defence budget of $26 billion represented a 5 per cent increase over inflation. Much of the increase was allocated to weapons procurement, but it still fell short of what was needed to keep the several major programmes on schedule. The Chirac budget also imposed small reductions in maintenance and military compensation. Overall, the Chirac government fell 0.2 per cent short of its 1986 campaign pledge to increase resources devoted to defence from 3.7 to 4.0 per cent of GNP.[22]

In sum, French conventional forces are generally designed for the short, intense combat seen for them under current French doctrine. Those forces would have to expand their personnel numbers and weapons stocks and improve their logistic support to survive in a more protracted conventional conflict. The creation of the *FAR* enables a more effective display of solidarity with the Allies in times of crisis and more flexibility as to where forces could be deployed. However, as French officials are quick to caution, there is no automaticity to the FAR's employment.[23] The 1983 Army reorganization moves more forces towards the north, making them somewhat better positioned to support the NORTHAG (Northern Army Group) area, where NATO needs greater depth, than as a reserve for CENTAG (Central Army Group). David Yost believes that 'France may be capable of contributing more conventional combat power in Europe than ever before during her membership in the Alliance',[24] but the critical question for NATO – and particularly for West Germany – remains how and when this military power would be employed.

THE IMPORTANCE OF FRENCH LOGISTICAL SUPPORT

Several officials and military analysts have noted that, while the availability of more capable French forces would significantly enhance NATO's conventional capabilities, the greatest contribution France could make in this regard would be to prepare and commit its logistical infrastructure to support of the overall Allied reinforcement effort.[25] This would be particularly true in the event of a protracted conflict. However, the current state of French logistic cooperation with NATO is such that many weeks of peacetime preparation would be required for France to give much support to Allied operations.[26] For France to provide timely and significant support to NATO operations would require even greater preparation and a willingness to confront the lingering domestic resistance to military plans that appears whenever there is talk of reintegrating France into the NATO military structure.

During the 1950s important LOC for CENTAG ran west to east through Central France to the Rhine. However, by 1963, economic, military and political considerations moved the US to rely more on the vulnerable Bremerhaven LOC system that had been established after the war, and the French LOC system was relegated to a secondary role in NATO planning. Nevertheless, France's move to take absolute control over access to its airspace and logistic infrastructure after the 1967 withdrawal from the integrated military command had a major impact on Allied planning. CENTAG LOC had to be redirected through Belgium and the Netherlands, and NATO lost the bulk of its rear area and a number of less risky resupply options, since French ports, railroads and highways are well developed and could be effective in supporting the NATO wartime resupply effort.[27]

Assured wartime access to French airspace and airfields would greatly increase NATO's options for airlift operations and aircraft dispersal. Even before 1967, Allied reliance on French airfields had diminished. President De Gaulle's ban on the storage of US tactical nuclear weapons in France after 1959 forced the US Air Force, Europe (USAFE), to redeploy nuclear-capable fighters to bases in the United Kingdom and West Germany. However, the total lack of access to French airfields after 1967 led to an increased density of aircraft deployments in West Germany, with attendant increases in vulnerability and strains on air traffic control and support facilities. As currently configured, French airfields are not equipped to service

NATO aircraft, generally do not have adequate munitions and fuel storage facilities for such a role and are already overcrowded with French aircraft.[28] Extensive NATO use of French airfields would, therefore, require the costly construction of new installations or conversion or enlargement of existing ones. Such a visible move towards peacetime cooperation in NATO would naturally require a major shift in French domestic political attitudes.

Thus a more realistic goal in the current political context would be for the NATO Allies to work with France in laying the groundwork for wartime access to French LOC. This could include peacetime testing of these networks and readying of stockpiles. Such actions could be taken discreetly, with adequate safeguards to ensure that French political sensibilities about implied guarantees are protected.

THE FRANCO-GERMAN DEFENCE DIALOGUE

The expanding Franco–German dialogue on defence is another avenue through which France can clarify its commitment to defend West Germany and one that appears to enjoy broad political support. This sort of bilateral reintegration into NATO defence planning allows Paris to maintain autonomy and avoid any appearance of subservience to US strategy. Moreover, this dialogue has become a dynamic force in the growing search for a European defence identity.

The Paris–Bonn dialogue was formally initiated at the October 1982 Summit between President Mitterrand and Chancellor Kohl.[29] However, its origins can be traced to earlier conversations between the two leaders' immediate predecessors.[30]

Paris initially viewed this dialogue as a means of counteracting what were seen by many French observers to be nationalist yearnings that would lead to neutralist policies in West Germany. Given that the expensive nuclear force modernization programme was constraining spending on conventional weapons and that there was lack in export orders, France also had new incentives to seek German capital for cooperative weapons development programmes. Both Paris and Bonn saw this dialogue as another mechanism to strengthen European solidarity in security policy. Indeed, after the poignant and historic meeting between Kohl and Mitterrand at Verdun, the relationship began to take on larger political significance, suggestive of the final reconciliation of the two former adversaries and of their common destiny in a new Europe.

Thus in 1982 the two parties agreed to activate a Defence Commission that had been established under the terms of the 1963 Elysée Treaty of Friendship. Three task forces, one on strategic issues, one on armaments collaboration and another on military cooperation, were also revived and have been meeting regularly since. These talks have become a major feature of relations between the two states with semi-annual summits, periodic meetings of defence and foreign ministers and Chiefs of Staff, and numerous lower-level contacts.

The impact of these discussions has already begun to be felt. The committee on strategic issues had reportedly discussed details about French nuclear targeting.[31] In 1984 senior West German officials, including former Chancellor Schmidt, began to ask for influence over any French decision to launch nuclear weapons against targets on German soil, including East Germany. As one top official said at that time, 'We want some sort of nuclear guarantee whereby France regards German territory, East and West, as its own security area'.[32] In late 1983, Gaullist leader Jacques Chirac floated a trial balloon by suggesting the possibility of appropriate German participation in an independent nuclear deterrent.[33] While this idea was quickly shelved, other French officials suggested that some understandings could be reached with West Germany on the nuclear targeting issue.

This dialogue has also touched on conventional military strategy and appears to be expanding. Helmut Schmidt's call for 'a major security initiative', whereby France and West Germany would jointly expand their reserve mobilization capabilities to the point that they could field 30 divisions, was greeted with silence in Paris.[34] Schmidt's argument that this force would be 'sufficient to defend the Western part of Europe and deter any attack' was not something Paris was willing to consider, given the nuclear orientation of French strategy and existing financial constraints. However, Schmidt suggested that if France were to extend her 'autonomous nuclear force to include protection to West Germany', then Germany should finance the rest of the programme, in part by savings from eliminating the nuclear role for certain dual-use weapons systems and by joint financing of conventional weapons.

French officials did not take up Schmidt's challenge initially, for fear of upsetting relations with the new conservative government in Bonn. However, in March 1987 after Kohl's re-election, President Mitterrand did invite Schmidt to the Elysée to discuss ideas for European defence.

Several prominent French defence analysts have advocated equally dramatic changes in their country's relationship with West Germany. Calling the non-belligerency option anachronistic, Pierre Lellouche has proposed an unambiguous French commitment to West German security.[35] Lellouche suggests that such a commitment could be provided by moving the 2nd Corps and the two new divisions of the *FAR* up to the front and positioning the 1st and 3rd Corps of the First Army on either side of the Rhine along with supporting tactical nuclear weapons. This action would not only double the number of French troops in the Federal Republic, but also bring French nuclear weapons and the bulk of France's armoured and anti-tank capabilities to German soil. French and West German tactical nuclear weapons would have to be harmonized in a bilateral nuclear planning group, which could be enlarged at some point to include the United Kingdom. Under this concept, France would still retain authority to order the use of tactical nuclear weapons after consultations with West Germany and would also reserve exclusive control over its strategic nuclear forces. Lellouche advocates negotiation of further technical understandings with NATO commands, but rejects the notion of any return to the Alliance's integrated military command.[36]

It is doubtful whether such a bold revision of French strategy and a firm French commitment to the defence of West Germany could be realized in the light of enduring political support for doctrinal flexibility and in the circumstances of adverse economic trends. France is in no financial position to embark on a major expansion of its general purpose forces, and most officials believe that deterrence is enhanced by maintaining uncertainty as to how France might protect her interests. Despite the possible political benefits of a larger troop presence, it would seem to make little sense militarily for France to place its most capable mobile forces so far forward where they would be more vulnerable to a Soviet short-warning attack.[37] Nevertheless, Lellouche's more radical policy proposals have touched off considerable debate and do reflect a response to shifts in public opinion.

In 1985, each of the major parties in France, except the communists, endorsed the concept of an 'enlarged sanctuary'. These moves mirror a dramatic shift in French public opinion.[38] In a mid-1985 *Le Monde* poll, 57 per cent said France should rush to aid Germany if she were 'seriously threatened' while only 19 per cent took the opposite view. Forty per cent said such an event would constitute a threat to 'the vital interests of France', and would therefore favour extending the nuclear guarantee to Germany: 24 per cent opposed the idea and 36 per cent abstained.[39] The pace and scope of Franco–German contacts continue to grow. A Mitterrand–Kohl summit on 18 July 1985 was devoted exclusively to defence issues. At that meeting, the two leaders agreed to set up a Bonn–Paris 'hotline'.[40] The largest joint manoeuvres ever held by the two countries – at the corps level – took place in 1986 and 1987, and joint training programmes for officers are being established. In 1987, units of the French *FAR* held exercises in West Germany.[41] In a historic agreement signed on 1 March 1986, Paris agreed to consult Bonn before using its tactical nuclear weapons on East or West German soil. In addition, French and West German military commanders were instructed under the terms of this pact to develop detailed plans for the employment of the *FAR* in the defence of West Germany. The 1987 agreement to form a 3000 man joint Franco–German brigade underlines this burgeoning degree of collaboration and potential commitment.[42]

All of this edges France somewhat closer to the concept of an enlarged sanctuary although full extension of the French nuclear umbrella over Germany is out of the question in the near term. France is not yet prepared to consider joint nuclear target planning with its German allies. Moreover, both countries recognize that a limited French shield could never substitute for the American nuclear guarantee. None the less, France has come a long way in clarifying its capabilities to intervene in a limited European War. The existence of the *FAR* and these other capabilities makes the French 'non-belligerency' option even more theoretical. Soviet planners might feel the need to neutralize this capability early in a conflict, bringing the war to France with or without a clear decision from Paris. Prime Minister Chirac is expected to continue to expand this military cooperation and to place greater emphasis on French conventional force modernization than did the previous Socialist Government.

THE LIMITED ROLE OF SPAIN

The implications of Spain's accession to NATO in June 1982 were cast in doubt four months later when the Socialist government, at best lukewarm about NATO, was swept into office promising a referendum on the membership issue. Since the March 1986 referendum came out in favour of continued membership, Premier Felipe Gonzales has stressed cooperation with NATO. An October 1986 government memorandum leaked to the press suggested that Spain was willing to contribute more to the common Western defence and to coordinate certain military tasks, such as naval operations, with NATO.[43] It appears that Spain now wants to play a constructive, albeit individualized, role in Western defence.

The Spanish have opted for a slightly modified variant of the French role in NATO. Like France, Spain is not a member of the integrated military command. However, unlike France, Spain is a member of the NATO Defence Planning and Military Committees and sends observers to the Nuclear Planning Group (NPG). Spain's situation is further complicated by its bilateral military relationship with the US, which was forged during the Franco era. The Socialist government has linked continuation of Spain's membership in NATO to substantial reductions of the 12 500 members of the US armed forces stationed in the country. Washington's negotiations with Madrid on extending its basing rights past the May 1988 expiration date have seen some sharp differences and the Gonzalez government has responded to pressure to obtain near-term American withdrawals.

Spanish officials have characterized their objective in all this as being to carve out a new role for the country in Western defence. They argue that the political and military situations in the past were such that the principal contribution Spain could offer to Western security was the US bases. However, it is argued, the commitment by all post-Franco governments to modernization of the military will enable Spain to play a more constructive role as an equal partner with the US and its other NATO Allies.[44] In practical terms, Spain sees the growth of its own capabilities as facilitating a reduction of the American presence and allowing for Spanish forces to assume new missions through joint planning and cooperation with NATO.

This raises the question of exactly what Spain could contribute to NATO's conventional defence given its current and programmed military capabilities and the facts of geography. With 320 000 active duty forces and a million reservists, the Spanish Armed Forces have a sizeable manpower pool. However, manpower costs, which in 1979 accounted for 60 per cent of defence spending, were trimmed to less than 50 per cent by 1986, and the government hopes to cut this to 40 per cent by reducing the number of active duty personnel. Moreover, most Spanish equipment, particularly for the Army, lags a generation behind that deployed with the principal NATO forces. The Spanish Army, which presently accounts for two-thirds of military manpower, has been trained primarily for internal security and territorial defence roles. The Air Force is designed for tactical support of ground and naval forces and has very limited power projection or transport capabilities.[45] The Spanish Navy has enjoyed fairly consistent growth and modernization over the past 25 years, and is in the process of acquiring a new light carrier and several new frigates.[46] The Spanish Navy has also had the most substantial operational experience with foreign services.[47] In addition, Spain has a flourishing armaments industry.

Given this very cursory review of Spain's military capabilities and interests, some general observations can be offered about its likely contributions to NATO's conventional defence. While some junior Army officers reportedly long for 'postings on the Rhine', Spanish ground forces will continue to have a national territorial defence role for the foreseeable future. At some point, it might be possible for Spain to contribute small ground and air units to ACE's Mobile Force (AMF) under some form of integration or an *ad hoc* arrangement,[48] Spain might also find it advantageous, as have the French, to participate in the integrated NATO air defence network (NADGE). The Spanish Navy seems most able to assume new operational responsibilities in NATO, perhaps expanding its patrol areas and anti-submarine warfare activities. Given the political problems likely to arise *vis-à-vis* Portugal and the UK over any adjustments to the command structure for the Iberian peninsula and adjacent waters if Spain were to be fully integrated into the Allied military structure, NATO planners might find it easier to negotiate separate arrangements with Spain on naval cooperation.

Somewhat akin to the situation with France, assured access to the Spanish military infrastructure would give NATO added depth and alternative, albeit lengthy, LOC that could prove useful in a protracted conventional conflict in the Central Region. Spanish airbases could be used by NATO for stockpiling equipment and as staging points for a variety of air and ground operations in Europe. Overall, however, such arrangements seem unlikely in the current political context.

Finally, Spain has been a valuable contributor to a number of cooperative armaments programmes and is an active member of the IEPG. Madrid has also expressed some interest in joining the WEU and playing a larger role in the development of a European defence identity. All of these tendencies would appear to redound to NATO's benefit, and should be encouraged. Spanish involvement in the Western defence debate and in the development of military equipment, even in a very particularized fashion, is preferable for both political and military reasons to the insularity of the past.

5 The Impact of New Technologies: Evolutionary or Revolutionary?

Several proposals for improving NATO's conventional forces have centred upon the potential contribution of a variety of technological developments that promise to revolutionize the effectiveness of non-nuclear weapons systems against a wide range of targets.[1] Referred to under the general shorthand rubric of 'emerging technologies', or 'ET', what is envisaged is the integration of a wide range of advances in munitions, precision guidance, sensor, data processing and communications technologies into complex high-capability weapons systems.

Critical support to such systems would be provided by rapidly evolving capabilities for all-weather, near-real-time surveillance and target acquisition at close range and up to 300 km behind the forward line of own troops (FLOT). In essence, these emerging technologies promise the capability for rapid direction and redirection of more lethal conventional firepower and more capable general purpose forces to those areas of the battlefield where they can be most effective. Moreover, a number of these new systems may be capable of accomplishing missions currently assigned to tactical nuclear weapons.

Concurrent with these developments has been a debate over the prudence of NATO's military doctrine and tactics discussed in the next chapter. This debate has been influenced by the growing appreciation of the capabilities of these emerging conventional technologies and mounting doubts about the utility of theatre nuclear weapons. There is no question that many of these emerging technology weapons would be more effective if employed in concert with revised military tactics. However, this debate over requisite conventional military capabilities and the appropriate doctrine for their employment has reopened old fissures within the Alliance.

While a number of the new weapons technologies have been tested and a few are operational, the bulk are still in the developmental stage or have not been integrated into weapons systems. Test data raise considerable doubts about the operational performance and reliability of many of these complex systems. Whichever technologies prove effective will have to be incorporated into national defence plans early on in the planning cycle. High costs are also likely to limit their procurement in militarily significant numbers given that little or no growth is projected in the defence budgets of most European countries.

American predominance in the development of most of the relevant technologies raises additional problems. While West Europeans fear that pursuit of a high-technology defence could mean concomitant dependence on American technology, the US – as noted earlier – worries that provision of certain advanced technologies to its European partners carries the risk of leakage or unauthorized transfer to the Warsaw Pact. There is concern in NATO that Soviet military counter-measures could erode the effectiveness of these new weapons capabilities or that a sustained diplomatic offensive by Moscow could undermine Allied commitment to force modernization. Thus formidable technological, political, military and economic hurdles will have to be overcome if NATO is to incorporate all or even some of these high-technology weapons in its defence. Because of these barriers, ET's impact on NATO's defence capabilities seems likely to be evolutionary rather than revolutionary, despite claims to the contrary.

THE NATURE OF THE TECHNOLOGIES AND THEIR APPLICATION

History is replete with examples of advances in technology which have had a profound impact on the nature of warfare. It has often been developments in a few specific areas that have had a dominant impact in any given period. The character of the First World War was markedly different in intensity from previous conflicts as a consequence of advances in chemistry and mechanical engineering which led respectively to the development of new, more potent explosives and to rapid fire weapons. Progress in aviation and the development of radar and other electronic wizardry allowed for unprecedented flexibility and precision in the targeting of forces during the Second World War.[2] Leaving aside the development and refinement of nuclear weapons, changes in the technologies for conventional land–

air warfare have hitherto been significant, but largely incremental, since 1945. However, emerging and accelerating developments in microelectronics, sensor technology and munitions could alter considerably the nature of non-nuclear conflict during the coming decade.

The driving force behind this change is the incorporation of advanced data-processing systems and a variety of optical, radar, infra-red and laser sensors into weapons and reconnaissance systems that makes possible extraordinarily accurate and timely target acquisition in all types of climatic conditions over a broad area. These new technologies are already having an impact on the capabilities of systems that NATO forces are procuring to meet the entire range of its military requirements. Emerging and programmed assets may enhance NATO's warning of, and ability to counter, an initial Warsaw Pact attack, as well as Alliance capabilities to diminish and defend against associated air power, to disrupt Warsaw Pact C^3I (command, control, communications and intelligence) – while maintaining the integrity of its own system – and to interdict aggressor reinforcement efforts. While emerging technologies are certain to have an impact on close combat, the most dramatic innovations could come if capabilities to detect, track, acquire with precision and attack with high lethality stationary and mobile targets as distant as 300 km forward of the line of contact are perfected.

These new capabilities for 'deep strike' have attracted considerable attention. Combined with adjustments in doctrine and tactics, such strikes could enable NATO, despite its quantitative military inferiority, to prevent a Warsaw Pact breakthrough of its forward defence without resort to nuclear weapons. This ability to acquire targets accurately at great depths, coupled with advances in munitions and stand-off weapons, may be particularly significant for some of the counter-air, reinforcement interdiction, and counter-C^3I tasks presently assigned to nuclear systems. None the less, a number of the technologies in question have applications in both close combat and deep strike missions.[3]

Microcomputers with digital and signal processing capabilities unimagined only a few years ago, and which are likely to expand exponentially in the very near term, are the nerve centres of an increasing number of conventional military systems. Advances in large-scale and very-large-scale integrated circuits (LSI and VLSI), which can carry out data-rate processing previously performed by mainframe computers, are already providing tactical surveillance systems with the capacity to track a large number of targets in near-real-time and pass relevant targeting data to battlefield commanders. As conventional military equipment grows more complex, microelectronic components play an increasingly important role in their operation. Microcomputers can be found in the guidance systems of most new ballistic and cruise missiles, in the fire control systems of new artillery, rocket launchers and tanks, and throughout the C^3I infrastructure.

The pace of innovation in microelectronics is no longer set primarily by military R&D efforts. Independent civilian technology development programmes in the West are increasingly yielding military spin-offs. The reversal of the traditional flow of technology in this field creates new political problems relating to control. Military research programmes continue to yield important new capabilities. For example, demonstration tests have been conducted on six new 'families' of smaller, less costly, and more reliable computer chips, developed under the US Defense Department's Very High Speed Integrated Circuit (VHSIC) programme. Insertion of VHSIC technology into operational weapons systems began in 1985. Development of second-generation VHSIC chips promises to yield hundred-fold improvements in processing power and greatly increase the performance of the whole range of military equipment using microelectronic components.[4]

Developments in surveillance and sensor technology are transforming all aspects of NATO's conventional military capabilities. A variety of new sensors on tactical reconnaissance platforms is providing NATO commanders with a capability for more precise all-weather day and night coverage of stationary and mobile Warsaw Pact forces at increasing distances. This tactical intelligence data is complemented by improved access to information from US national reconnaissance satellites and other space assets. For example, data from the Global Positioning System (GPS) on the US NAVSTAR satellites will, when the full constellation of satellites is in place in 1988, provide US pilots with precise three-dimensional location data that will enable pinpoint targeting anywhere in the world. GPS receivers could even be incorporated into the terminal guidance systems of weapons to provide near-perfect accuracy to 'fire-and-forget', long-range, stand-off weapons.[5] There is some resistance in the US intelligence community to the notion of passing more sensitive information to both US and other NATO commanders in the field. It is feared that making such data more widely available increases the risks that invaluable national collection systems could be compromised. However, there seems no reason why secure means of data transmission and storage cannot be developed to ensure that Allied operational commanders have the benefits of the full panoply of US intelligence and other space assets.

New sensors include infra-red devices which can detect vehicles at night or under camouflage, synthetic aperture radars which can produce photo-like images through cloud cover and detect movement of aircraft and troop convoys, and electronic intelligence sensors for precision location of hostile air defence radars and other electro-magnetic emitters. These collection systems are becoming more interactive as they are linked together by computers at tactical information fusion centres. Thus information from one system can be used to alert or tailor the collection activities of another more efficiently.

One of the most significant new tactical reconnaissance and surveillance platforms in production is the US TR-1 high-altitude, long-endurance aircraft. A derivative of the U-2R, the TR-1 was procured to support the Tactical Reconnaissance System (TRS) (a combination of active radars and passive surveillance sensors) as well as the Advanced Synthetic Aperture Radar (ASARS) and the Precision Location Strike System (PLSS). ASARS can provide radar images of ground targets such as tanks and parked aircraft at ranges in excess of 150 km. PLSS was designed to yield extremely accurate real-time targeting data on enemy signal emitters (radars, jammers and communication systems), and can direct aircraft and missile strikes against them. However, during the PLSS prototype tests it was found that processing and analyzing the vast numbers of signals collected during fast-paced combat operations is far more difficult than expected, and the programme was cancelled. In addition, the TRS prototype tests revealed several problems that will require modification, resulting in several years' delay in this TR-1 sensor package.[6]

Development work is under way in the US on a new long-range radar surveillance system that would be the critical support element for non-nuclear interdiction of second-echelon Warsaw Pact forces. The US Army/Air Force Joint Surveillance and Target Attack Radar Systems (JSTARS) programme, established in 1982, is designed to provide both services with capabilities, which they currently lack, for real-time monitoring and identification of both stationary and moving ground force targets at long ranges (50 km and more) and over a wide field of view. The This airborne system would transmit the data it collected to a remote truck-mounted ground station for distribution to various weapons systems. JSTARS may also have an autonomous capability to direct weapons strikes against identified targets.

Differing Army and Air Force mission requirements and platform preferences, coupled with developmental problems, have slowed JSTARS considerably. The Army wanted to deploy JSTARS on its small *Mohawk* OV-1 reconnaissance aircraft. The Air Force debated internally and with Congress as to whether JSTARS should be deployed on the TR-1 or on the current platform, the E-8A, a military version of the Boeing 707 airliner, where there will be ample room for on-board processing capabilities as a substitute for ground stations. During generally disappointing tests of the *Pave Mover* radar (a simplified precursor of JSTARS) the system nevertheless monitored a moving tank column and guided F-4E aircraft for 75 miles to conduct a successful mock attack on the tanks, which the pilot could not see, by feeding data directly into the F-4's computers.[7] JSTARS, now in full-scale development, is expected to improve the management of NATO battlefield interdiction efforts and stand-off interdiction attacks. The system is scheduled to begin prototype testing during the late 1980s, with deployment of ten aircraft envisaged in the 1990s.

The French and British are developing two less sophisticated surveillance programmes. The British *Astor* programme has experienced developmental problems, and neither a manufacturer nor a platform – *Islander* and *Canberra* aircraft are candidates – has been selected. The French are working on *Orchidée*, which is a helicopter-mounted surveillance system similar to the US Army's original design for JSTARS.[8]

The 18 E-3A AWACS aircraft now deployed with multinational NATO crews in Europe can provide unique early warning and unprecedented air battle management capabilities. These AWACS aircraft, and a force of six to eight more E-3A aircraft with maritime and aerial surveillance capabilities to be purchased by the United Kingdom, will form the NATO Airborne Early Warning Force (NAEWF). France will also purchase three to five E-3As in an arrangement with Britain whereby the two countries will pool training and maintenance activities, thereby reducing costs. The initial French and British systems are scheduled to enter service by 1991.[9] As few as three E-3As flying overlapping orbits in central

Europe can provide high- and low-altitude surveillance of aircraft operations from the North to the Mediterranean Seas. Flying in their normal operational configuration 100 nautical miles within NATO borders, these aircraft can provide coverage of all low-level attack corridors, spotting ground-hugging Warsaw Pact aircraft, which cannot be detected by land-based radars, shortly after take-off. AWACS controllers can utilize this tracking data to manage several aspects of the NATO air battle.

The AWACS can vector NATO interceptors against all types of attacking Warsaw Pact aircraft including those hunting the AWACS aircraft itself. A similar capability was demonstrated, albeit in a less dense combat environment, by the Israeli Air Force during its 1982 invasion of Lebanon. The radar data from an E-2C *Hawkeye* was used with great success to vector Israeli F-15s and *Kfirs* against Syrian MiGs.[10]

The AWACS may also be capable of transmitting this target-tracking data in real time to NATO surface-to-air missile (SAM) sites, enabling a more efficient utilization of air defence firepower. In addition, AWACS is expected to be capable of warning NATO close-support and interdiction aircraft of the whereabouts of Warsaw Pact interceptors and informing NATO commanders where hostile aircraft that survived the initial combat action are landing.

This growing wealth of tactical information has already created new problems for data handling and analysis. NATO fusion centres are already suffering from information overload that slows processing. In a crisis this situation would be compounded, precluding timely transmission of target acquisition data to users. Thus NATO needs to tailor reconnaissance efforts to target acquisition requirements and must develop ways to filter out the most critical information. At the same time NATO needs a widely deployed, secure, jam-resistant and interoperable communications capability to ensure a smooth flow of information from collection systems to processing centres and on to consumers. Progress has been made on this score with the programmed incorporation of the Joint Tactical Information Distribution System (JTIDS), a secure digital and voice communications system, into the AWACS, many NATO air defence sites, and a number of tactical aircraft. The first successful flight test of a JTIDS terminal was conducted in 1985 using an F15 aircraft.[11]

Given the important role that air- and ground-based early-warning and battle management systems would play in NATO defences, they would obviously be prime targets of Warsaw Pact attacks and electronic warfare measures. Indeed, the Achilles heel of the high-technology conventional battlefield is the C^3I system that would be employed in its management. Development of a C^3I system with sufficient protective measures or redundancies to survive in what would be an extraordinarily intense combat environment will be a major challenge for NATO. The Soviet Union has a well-developed radio-electronic combat doctrine which emphasizes systematic disruption of the opponent's C^3I system by attacking it with a combination of appropriate lethal and non-lethal means as the situation warrants.[12] Clearly the limited number of fixed command posts where fusion of all NATO's tactical intelligence data and battle management takes place is a major vulnerability of the current system. Mobility is one approach to limiting system disruption, and NATO is planning to procure land-mobile command posts as well as early-warning and battle-management aircraft. None the less, even a system like AWACS – with its mobility, ability to manage fighters in its own defence, and impressive electronic counter-measure system – remains vulnerable to long-range SAM and air-to-air missiles while airborne, and to a variety of other threats, including tactical ballistic missiles and *Spetznaz* forces, while on the ground.

The growing complexity of NATO's C^3I system makes it more vulnerable to disruption as a consequence of incapacitation of various nodes. Thus the detailed information on the activities of advancing Soviet armoured columns collected by a TR-1 aircraft would be useless to battlefield commanders if the TR-1's cooperating processing station on the ground were destroyed by hostile fire or if communication links were jammed. The search for ways to cope with these vulnerabilities has included an examination of the utility of expendable, unmanned aerial surveillance and targeting platforms and the feasibility of greater exploitation of space-based assets, both with direct communications links to battlefield commanders.

The Israelis demonstrated how the integration of unmanned surveillance and targeting platforms with aircraft and missile strikes could be achieved during their June 1982 invasion of Lebanon.[13] The Israelis used unmanned *Scout* remotely-piloted vehicles (RPV), which are equipped with an electro-optical sensor package and a digital data link, to provide various echelons of the command structure with imagery of Syrian airfields and SAM sites in near real-time. Inexpensive unguided drones which can simulate fighter aircraft electronically were deployed to keep Syrian SAM target acquisition radars radiating while various missiles with radar-homing sensors were launched against them.

NATO has deployed a number of RPV over the past decade including the Canadian AN-USD 510, the French R-26 and the Belgian R-26. Payload limitations and sensor size have limited these vehicles to infra-red detectors and optical imaging.

American advances in senson and guidance system miniaturization and small airframe technology have greatly improved RPV capabilities in recent years. The Army has been having considerable success using the existing *Skyeye* RPV for reconnaissance missions in Central America.[14] US programmes for sophisticated new RPVs, the Army's MQM-105 *Aquila*, and the Air Force's *Pave Tiger*, have however seen tremendous cost-growth and reduced capabilities in their development cycles. *Aquila* was developed to provide real-time, all-weather reconnaissance and target acquisition data to artillery batteries in high-intensity combat environments. It was to be fitted with a daylight television camera, laser range-finder, a night/adverse weather forward-looking infra-red system (FLIR), and a sensitive electronic counter-measures suite. Now *Aquila's* electronic warfare capabilities have been scaled back, and its night capability will not be realized for several years. Nevertheless, flight testing was completed in 1987. Procurement of up to 376 *Aquila* systems, costing over $2 billion, is envisaged. At the same time, the US Army is planning to procure a lower-cost general purpose RPV, called by the acronym IEWUAV, to perform corps-level intelligence and electronic warfare missions in low-intensity combat environments. The Army plans to procure a mix of *Aquila* and IEWUAV beginning in 1990.

The US has been expanding its utilization of highly survivable satellite command, control and communications systems such as DSCS III and MILSTAR. As noted earlier, efforts are also under way to provide tactical commanders with more data from various national reconnaissance systems. It must be assumed that the high-technology battlefield of the 1990s will be one where reconnaissance support and C^3 could be periodically interrupted or partially degraded by hostile action. However, it is most unlikely that electronic warfare and other measures could totally incapacitate these critical support elements. While the flow of data and the performance of various sensors will not always be up to design specifications, the impact of these new electronic support capabilities will be considerable. In the face of disruption, *ad hoc* arrangements to patch together the C^{3}I system may well restore some of its integrity. None the less, despite these technological innovations, NATO's C^{3}I capabilities are likely to remain sub-optimal without a greater degree of integration among national components and across service lines. Fortunately, most of the new high-technology weapons are not entirely dependent on centralized C^{3}I support.

Developments in Munitions

Many of the emerging weapons capabilities trace their genesis to advances in technologies other than the fabrication of silicon chips. There have been dramatic developments in conventional munitions and in rocket and missile guidance technologies, some of which have already proved their effectiveness in combat as well as on the test range. These weapons have applications both in close combat and in deep strikes. A number of European as well as American firms are in various stages of developing mid-flight and terminal-guidance packages for a wide range of delivery systems.

Advances in the technology of small explosive devices have led to the design of area impact and precision-guided submunitions that multiply and disperse the effects of a single weapons payload delivered to a target. For example, the warhead fitted on a single rocket of the MLRS which is now entering service in several NATO armies can disperse 644 M-77 submunitions, each with an explosive force equivalent to a hand grenade.[15] The charge is shaped in a way that gives it the capability to penetrate light armour. The US *Gator* mine system (CBU-89/B), scheduled to enter service in the late 1980s, will provide a relatively simple target-activated submunition package. *Gator*, designed to be carried on aircraft in the Air Force's standard 1000-pound Tactical Munition Dispenser (TMD), is a mix of 72 anti-armour and 22 anti-personnel mines which can be scattered over a wide area to create instant minefields. The *Gator* munitions are 'target-activated', meaning that they detonate when troops pass by or a vehicle rolls over them. *Gator* is sufficiently intelligent to distinguish between valid and false targets and to detonate only when the target comes within its lethal range.[16]

The potential increase in the effectiveness of conventional firepower afforded by the dispersal of lethal submunitions is best illustrated by projections developed in a 1982 West German study chaired by Manfred Woerner and Peter-Kurt Wurzbach. Using existing weapons, the destruction of a Soviet breakthrough group (assumed to comprise of some 600 tanks, 500 armoured vehicles, 50 artillery batteries, 200 SAM and 300 wheeled vehicles) was calculated to require 33 000 tons of gravity bombs delivered by 5500 aircraft sorties. The same mission might some day be accomplished by 600 aircraft sorties dispersing only 3000 tons of unguided submunitions or possibly by as few as 50–100 sorties delivering terminally-guided submunitions.[17] The study estimates that one ton of advanced technology submunitions could achieve the same effectiveness against a company as a 2–3 kiloton nuclear weapon. This report also concluded that to destroy 60 per cent of a Soviet division, thereby rendering it ineffective, with existing non-nuclear weapons would require 2200 aircraft sorties or 10 000 missiles. If unguided anti-armour submunitions were used, the same destruction could be achieved with 300 sorties or 1500 missiles. Guided submunitions would reduce the required number of missiles to 50–60. It would probably take 20–25 10-kiloton nuclear weapons to inflict this kind of damage.

Two prototypes of 'smart' anti-tank submunitions were tested, albeit with disappointing results and under artificial conditions, during the US Defense Advanced Research Project Agency's (DARPA) $275 million *Assault Breaker* programme.[18] Both test payloads were carried to and dispersed over the target area by short-range ballistic missiles. A prototype JSTARS radar was used on some of the flights and attempts were made – with only limited success – to have it issue mid-course commands to direct *Assault Breaker* missiles towards targets.

One of the systems tested in the *Assault Breaker* programme was a package of terminally guided submunitions (TGSM). After ejection from the delivery vehicle, infra-red sensors on each of these submunitions were activated at an altitude of about 1 km and began searching for the heat signatures of the target tanks. A guidance package adjusted the missile's tail fins and the munition's trajectory to ensure target acquisition, at which point a shaped charge was ejected to pierce through weak points in the tanks' armour. After ten failures, one of these terminally-guided submunitions packages managed – under optimal environmental conditions – to destroy five stationary M-47 tanks. However, this guidance success has to be qualified because the M-47's exhaust system runs on top of the rear deck, giving it a relatively high infra-red signature, particularly compared to the Soviet T-62 tank which has a side exhaust. Moreover, the system was tested in the early morning in the cold desert when the thermal contrasts between the desert and the hot tanks would be greatest. DARPA's own report on the programme concluded that a more sophisticated guidance system will be necessary to destroy a typical Soviet tank.[19]

The other system tested was the *Skeet* self-forging fragment. Pairs of *Skeets* were dispersed from a special delivery vehicle at a predetermined height. Each *Skeet* has an infra-red guidance system that is designed to seek out targets from low altitudes and after landing on the ground. When a target is acquired, an explosive charge is triggered which forms an armour-piercing molten metal projectile out from the body of the *Skeet*. This is directed towards the target. However, *Skeet* failed to hit any of the targets during DARPA's *Assault Breaker* test programme.

Due to the numerous failures and the contrived conditions of the one successful test, doubts remain about this technology. New testing programmes are underway. It is unclear how effective these 'smart' munitions would be against moving targets or under conditions such as bad weather, defensive counter-measures and terrain 'clutter' that one would expect to encounter on a West European battlefield. In fairness, however, it should be noted that this is a relatively immature technology which could be perfected over time. The other components of the *Assault Breaker* deep strike system, a simplified JSTARS radar and two missile systems, performed fairly well in tests but successful integrated operations of the entire system have yet to be demonstrated. The *Assault Breaker* programme also illustrated a weakness common to a number of these new high-technology weapons: overall system performance can be critically degraded by the malfunctioning of any one of its several components.

The two missiles involved in the *Assault Breaker* demonstration programme – the T-16, a derivative of the *Patriot*, and the T-22, a variant of the *Lance* – were the principal contenders for the Joint US Army/Air Force Tactical Missile System (JTACMS). Originally two separate programmes, JTACMS was to become both the principal second echelon interdiction weapon for the corps commander and the Air Force's conventional stand-off weapon for use against heavily-defended targets. However, the two services have quite different target priorities and, while the Army favoured a ballistic missile, the Air Force preferred a cruise missile. As a result, in May 1984, the two services decided to develop separate but theoretically compatible sub-systems, based on a common agreement concerning missions and targets.

The Army now intends to procure an extended-range version of the MLRS known as the ATACMS. The ATACMS will be capable of reaching targets up to 70 km away and can be fired from existing MLRS launchers. While ATACMS will have much better accuracy than *Lance*, it will not have a terminal or mid-course guidance system, which would be required to achieve a precise hit on a point target. The Air Force is interested in a 550 km-range air-to-surface cruise missile. ATACMS is expected to carry anti-tank and anti-personnel submunitions, but it could also disperse anti-tank mines. ATACMS would be targeted on follow-on forces, air defence systems, and command and control facilities. The pace of the programme has been slowed by problems of incorporating the capabilities desired by the two services and by Congressional concern about mounting costs. The system, which is a critical component of both the US Army's and NATO's concept of an extended battlefield, is not now scheduled to begin operational testing until FY 1989 and seems unlikely to be procured in a standardized form.

New Counter-Air Capabilities

A number of weapons programmes incorporate precision guidance or novel munitions technologies that have the potential to improve NATO counter-air capabilities. Four basic types of munitions are required for conventional air attack against airbases: cratering munitions to disrupt runways and taxiways; area effects fragmentation munitions to destroy unsheltered aircraft and equipment; area denial mines to restrict movement of personnel, equipment and aircraft; and weapons to destroy protected aircraft. A number of these types of systems have been successfully tested and some are being deployed.

The French *Durandal* airfield attack system is designed for direct delivery by high-speed low-altitude aircraft. It is a rocket-boosted bomb that penetrates deep into runways before exploding, thereby maximizing cratering. The US Air Force has deployed *Durandal* on F-111s, which can carry up to 12 of these bombs. Two similar cratering munitions are being deployed on NATO tactical aircraft, the British SG-357, delivered by *Tornado* aircraft fitted with the JP-233 dispenser, and the West Germany STABO, which is carried by *Tornado* using the MW-1 dispenser. The US is presently working on a more advanced cratering munition called the Boosted Kinetic Energy Penetrator (BKEP) which, while smaller than *Durandal*, may be even more effective.[20]

At least two types of airfield denial mines are also being deployed for aircraft delivery, the British HB-876 and the German MUSPA. In addition a wide variety of fragmentation munitions are already available for damaging unsheltered aircraft, such as the US Combined Effects Munition (CEM) and the West German MUSA bomblet.

Shelter attack weapons have yet to be perfected, although a number of R&D programmes are under way. The West Germans are currently developing an anti-shelter weapon, but more work needs to be done in this area of technology.

There has been considerable interest in developing guided stand-off munitions dispensers for NATO aircraft and utilizing surface-to-surface missiles to deliver airfield attack munitions packages because of the density and increasing sophistication of Warsaw Pact air defences. Several ballistic and cruise missile systems equipped with submunitions are being considered by the US Air Force for stand-off counter-air missions at ranges from 20 to 350 nautical miles. One programme option, now cancelled, was the Medium-Range Air-to-Surface Missile (MRASM), a 200-mile version of the *Tomahawk* cruise missile that could deliver a 1000 pound submunition payload, such as 28 BKEP.

Longer-range missile systems with greater payload capacities have also been examined for delivery of these advanced conventional munitions payloads against both airfields and other targets. The system most often discussed is the CAM-40, a 40-inch solid rocket booster, which is a variant of the *Pershing* Ia or the *Pershing* II first stage. From launch sites in West Germany the CAM-40, fitted with a sophisticated guidance system, could deliver a 3000 lb payload of improved conventional munitions (ICM) against the 105 most threatening Main Operating Bases (MOB) and Dispersed Operating Bases (DOB), while a two-stage variant could strike the remaining 36 important Pact airbases in Eastern Europe. The European Security Study (ESECS) estimated that about three CAM-40 missiles would be required to shut down a runway for three days, which would suggest a total requirement of at least 900 missiles to target one runway and one taxiway at each of the 141 MOB and DOB.[21] This large missile requirement would necessitate a larger booster to deliver two or three times the payload of the CAM-40. A US concept that was scrutinized was the BOSS/AXE which envisaged the use of the *Trident* C-4 booster to deliver a 14 000 lb payload to ranges in excess of 350 nautical miles. The ESECS study estimated that only one BOSS/AXE with kinetic energy penetrators would be required to destroy a runway. The ESA's *Ariane* booster has also been considered as a delivery system for a large payload of ICM. An even larger missile/warhead concept that has been looked at is the Total Air Base Attack System (TABAS).

Utilization of these and other stand-off weapons against the 73 Warsaw Pact air force MOB while flight operations are under way could greatly reduce the air threat to NATO. Shutting down the principal airbases would force Pact air forces to use the 68 less well protected DOB. Full integration of AWACS data into the fire control systems of these weapons might enable specific targeting of those airfields to which hostile aircraft are returning following initial engagements. Even disruption of activity at these airfields could be a forceful blow to Warsaw Pact aircraft operations given the East's limited vertical and/or short take-off and landing (V/STOL) capabilities. Effective attacks against the MOB would reduce the number and pace of sorties so critical to the success of Warsaw Pact offensive air operations. However, the future scope for the use of larger ballistic missiles for delivery of ICM payloads is uncertain. In addition to some operational problems, the INF Treaty precludes the deployment of ground-launched missiles with ranges of 500–5000 kilometres.

The ESECS study suggested two primary goals for attacks against the MOB. First, strikes should be conducted against runways and taxiways to halt sortie generation, which would essentially freeze Warsaw Pact aircraft at their bases and require those already airborne to return to less well equipped and defended DOB.[22] Second, follow-on attacks could be launched to destroy important targets at the MOB, such as sheltered aircraft and maintenance facilities, thereby diminishing sortie generation capabilities over the long term. In addition, the DOB could be attacked in this second round, with the aim of closing runways and taxiways and destroying aircraft on the ground. Under this concept, manned aircraft, preferably employing stand-off weapons, would be used to follow up and augment the initial missile attacks.

Air Defences

The other critical element in the counter-air equation is NATO's ability to defend its own airfields. To be able to generate sorties for sustained air combat and ground support, NATO must be able to protect airfields for the launch and recovery of aircraft. NATO is undertaking major efforts to improve its air defences, and France participates in these efforts through the NADGE. A new NATO plan for European air defences has been developed, under which all obsolete nuclear-armed *Nike* systems and some *Hawk* units in the Central Region (save for Belgian *Nike* systems which are to be withdrawn without replacement) will be replaced by the non/nuclear *Patriot* system. The *Patriot* system integrates an advanced all-altitude air defence missile with a phased-array radar that has a high resistance to electronic counter-measures and the ability to direct several missiles to their targets virtually simultaneously. *Patriot* entered service with US Army units in Europe in late 1984, and several allied countries are considering its procurement for their forces. Currently-deployed *Hawk* systems are being upgraded with new missile components and an improved guidance system that will enable the simultaneous direction of several missiles to their targets.

In addition, considerable progress has been made on joint European-American arrangements for improving point and area air defences at US air bases. The US Air Force is procuring the *Rapier* missile to defend bases in the UK. Similarly, in 1984 the US and West Germany finally agreed on the terms of defence arrangements for US air bases in West Germany. Since 1986, the US sold 14 *Patriot* fire units to West Germany, with another 12 US-owned *Patriot* fire units being manned and operated by the Germans until the year 2000. The Germans will also buy 95 Euromissile *Roland* fire units, 27 of which will be deployed to defend three US air bases.[23]

Advanced technology under consideration includes a mobile air defence weapon and anti-ballistic missile (ABM) systems and there is increasing debate as to whether NATO needs to develop an anti-tactical ballistic missile (ATBM) system. The new generation of more accurate Soviet theatre missiles SS-21, SS-12 mod and SS-23, which can be fitted with improved conventional, as well as nuclear and chemical, warheads poses an emerging threat to NATO airbases and rear areas (although the SS-12 and -23 are now to be eliminated under the INF Agreement).[24] Moreover, the Soviet Union is soon expected to deploy the SA-X-12, which may be capable of engaging NATO's *Lance* (and *Pershing* Ia and II missiles until these too vanish from the inventory under the INF deal) as well as aircraft at all altitudes and cruise missiles.[25] While a number of ongoing research programmes under the rubric of the US SDI, particularly in the area of terminal defence, may have relevance to the development of a theatre ATBM capability, attention has focused on the question of a *Patriot* upgrade.[26]

When the *Patriot* programme began in 1965 under the name SAM-D, it was expected to have certain capabilities against tactical ballistic missiles. However, because of technical doubts about radar capabilities and concerns that development of ATBM by the US and the USSR would undermine the ABM Treaty, this ABM capability was not expressly pursued in the development of the system. However, at the April 1984 meeting of the NATO NPG, US Secretary of Defense Caspar Weinberger reportedly informed his counterparts that an upgraded *Patriot* system could serve as an ATBM layer in the extension of a strategic defence umbrella over Europe. Raytheon, builder of the *Patriot*, is developing new software for the system's surveillance radar which would make it capable of engaging ballistic missile re-entry vehicles, but this will have the side effect of reducing its air defence capability.[27] The Defense Department is considering the installation of switches on each *Patriot* fire unit radar to enable it to change from missile to air defence modes. Further improvements to the *Patriot*'s software, guidance system and warhead have also been considered. An improved *Patriot* would be intended to provide NATO air defence sites, airfields and theatre nuclear forces with defence against upgraded Soviet short-range ballistic missiles (SRBM), which might be used in the opening phases of any conflict in Europe as a prelude to a full aerial assault. Raytheon is also working on hardware modifications for the *Improved Hawk* system that would enhance its capability to attack cruise missiles such as the AS-15, which is similar to the US *Tomahawk*.

A number of more exotic technologies, such as ground-based lasers and rail guns, are being explored for ATBM and other air defence roles. West German Defence Minister Manfred Woerner has become a leading proponent of a so-called European Defence Initiative (EDI) for Extended Air Defence (EAD) which would develop some of these advanced technologies for ATBM.[28] This notion of an EAD has attracted growing support as a result of concern about the Soviet conventional missile threat and about the need for a European high-technology defence initiative to keep pace with the US SDI. Two German firms are collaborating on the development of a high energy laser (HEL) for use against low-level air threats.[29]

Systems for the Close-In Battle

NATO plans or is considering the deployment of a wide range of weapons systems to meet the challenges of forward defence, such as stopping the advance of armoured units and suppressing Warsaw Pact artillery and rocket fire. The first units of the MLRS have entered service with the US Army in Europe and will be built and fielded by five other members of the Alliance starting in 1989. MLRS offers a significant increase in firepower and important gains in range over existing artillery. As currently configured, each MLRS launcher can fire 12 surface-to-surface rockets within one minute at ranges out to 30 km. It would require at least three 8-inch artillery battalions firing simultaneously to deliver a weight of fire equal to the thousands of submunitions that one salvo of 12 MLRS rockets delivers. This kind of intense fire would be necessary to suppress the massive barrages of artillery fire that would accompany any Warsaw Pact offensive in Central Europe. MLRS is seen as an important supplement to traditional tube artillery. In Phase 1, MLRS will be fitted with M-77 bomblets, and in Phase 2 it will be deployed with the German AT-2 mines. Armed with terminally-guided submunitions using millimetric radar guidance in Phase 3, currently expected in the mid-1990s, MLRS could also be effective against armoured assaults.[30] The millimetric radar is expected to suffer less degradation in the face of counter-measures and other environmental factors than the infra-red sensors used in the *Assault Breaker* programme.

Guided artillery munitions is another area where there have been considerable technological advances. Over the past decade the US Army has developed a laser-guided artillery shell known as *Copperhead*. This system is entering service but, while certain development problems have been overcome, others persist. For example, the *Copperhead* is now capable of accurate fire in smoke and fog 75 per cent of the time, but heavy snow or rain will still degrade performance.[31] Moreover, for all its technological sophistication, *Copperhead* is still dependent on two forward observers who use a laser target designator and must radio target coordinates to the firing battery. Not only is the laser device cumbersome for soldiers to deploy, but the laser must be held on the target while *Copperhead* is in flight. Soviet tanks equipped with laser detection devices would be able to ascertain that they were being illuminated and could engage the forward observers or deploy self-screening smoke. These potential deficiencies have led the Army to develop an RPV to locate and designate targets for laser-guided weapons.

NATO is developing and deploying a wide array of anti-armour precision-guided munitions (PGM) that can be fired from helicopters, vehicles, artillery and by individual soldiers. Many of these PGM incorporate laser target designators and guidance systems as well as advanced shaped charges and other armour-penetrating devices that increase their lethality. Most current anti-tank guided missiles (ATGM), such as the long-range *TOW* and *HOT* and short-range *Dragon* and *Milan*, utilize command to line-of-sight guidance, necessitating continuous target tracking by the operator until impact. However, the third-generation fire-and-forget PGM under development in the West appear likely to overcome several of the significant drawbacks of earlier versions, such as operator vulnerability and weather limitations.[32] These systems will build on advances in millimetric-wave radar and infra-red sensors configured in focal-plane arrays. The resolution of these arrays is such that they can distinguish between vehicles of different sizes, thereby permitting selective targeting. European firms have been particularly active in developing new types of anti-tank guided weapons (ATGW). In addition to multilateral work on the third generation *HOT* and *Milan*, Great Britain is developing the *LAW-80* man-portable, unguided anti-tank weapon for short-range attack of main battle tanks.

The debate as to whether PGM favour the defence over the offence or can wholly compensate for the quantitative inferiority of NATO's armoured forces in the Central Region is unlikely ever to be resolved.[33] It is evident that the third generation of these weapons will be more lethal and reliable than their predecessors. Developments such as Soviet reactive armour may blunt somewhat the effectiveness of infantry-fired ATGM and other shaped charge munitions. However, the US Army has also developed a fairly inexpensive modification of the *TOW* missile that will enable it to defeat reactive armour.[34]

CONCLUSION: ET's UNCERTAIN FUTURE

As all these technological and military uncertainties show, there is no 'magic bullet' that will solve overnight all the shortcomings of NATO's conventional military forces. Most of the new technologies which promise to correct various problems or enhance certain capabilities are unlikely to be deployed in significant numbers until the end of the decade, and they may well turn out to be very expensive. For example, the ESECS study estimated that acquisition of a force of 1000 short-range MLRS systems to deal with first echelon forces, 5000 stand-off cruise missiles with 'smart' submunitions to attack follow-on forces, and 900 conventionally armed ballistic missiles for counter-air operations could be procured at a cost of $22.5–30 billion over a decade and would provide NATO with a solid conventional defence.[35] Other estimates for such systems vary widely, and it is premature to advance specific cost figures with much confidence at this point.

The question of emerging technologies attained a new degree of political saliency after the June 1982 meeting of NATO defence ministers. At that meeting US Defense Secretary Caspar Weinberger presented a paper outlining four areas where new technologies could be exploited in the hopes of greatly improving the capabilities of NATO's conventional ground and air forces. This proposal was referred to the NATO CNAD and other bodies for further study. While advanced quite cautiously, the ET Initiative – as it came to be called – received a frosty reception from most European ministers who found it a dubious path to enhancing conventional deterrence and one which was certain to lead to a 'one-way street' where they would be absorbing – and paying – for such technology as the US would be willing to share under strict export controls.[36] Real progress on the programme outlined in the Weinberger initiative will not be possible unless something is done to make it more attractive to European governments. Co-production arrangements could offset part of the European trade imbalance and employment losses, but in the long run these measures will not be totally satisfying. The European members of NATO will not want to perpetuate their technological lag in critical areas simply by producing American innovations under licence. Furthermore the European nations themselves have a number of new weapons to offer, primarily those with applications for conduct of the close-in battle, some of which are at least comparable to US systems.

In April 1984 the CNAD selected four principal areas of effort, involving 11 large defence projects, as the initial programmes for the exploitation of emerging technologies.[37] Each of the 11 projects was selected because it or a similar programme was already in the defence plans of at least four countries. The individual Armament

Directors were charged with the task of enlisting their government's participation in one or more of these projects. The majority of the 11 programmes are most appropriate to support the close-in battle, although a stand-off surveillance system is also included in the package.

Concern about the twin American challenge posed by SDI and the ET initiative might motivate West European governments to overcome existing obstacles to developing a more competitive European technological base. They might agree to pool resources and personnel in the defence and basic research sectors. *Eureka* and other mechanisms for European technological cooperation have made some initial progress along these lines, but major issues have still to be resolved. The difficult history of the EFA cooperative fighter development project illustrates clearly how differing national economic and military priorities can get in the way of cooperative weapons development efforts.

Thus it may be a misnomer to speak of a technological revolution on the battlefield. NATO forces are likely to realize these dramatic increases in capability only in an incremental fashion. Moreover, the impact of technology cannot be considered in isolation. There are many other factors – such as strategy, doctrine, training, readiness and morale – which must also be considered in any appraisal of the evolving military situation in Europe.

6 Changes in Concepts and Tactics

NATO's military tactics and basic operational concepts have also come under scrutiny in the search for a more effective conventional posture. Indeed, some analysts argue that the focus on more weapons and sophisticated technology is mistaken because dramatic improvements in NATO's conventional capabilities can only be achieved by operational innovation or major changes in strategy. In the middle of the spectrum in this debate are those, such as the most recent SACEUR, General Bernard Rogers, who argue that the defensive Allied strategy of Flexible Response and Forward Defence remain fundamentally sound tenets of Western deterrence which can be made more effective by operational adjustments.[1] Indeed, as General Rogers has suggested, achievement of the current NATO force goals would enhance Allied capabilities to a point that would reduce the likelihood of NATO's early use of nuclear weapons in a conflict. Others, who generally agree that NATO's current strategy is sound, argue that broader changes in force structure and plans (particularly for very deep strikes into Warsaw Pact territory) are required because of advances in Soviet nuclear and conventional capabilities.[2]

This chapter explores the genesis of these military concepts and assesses their potential to enhance Western deterrence.

DEEP STRIKE CONCEPTS

Coincident with and influenced by the advances in technology described in the previous chapter, there have been a number of significant developments in NATO and US military planning. A common feature of the new operational concepts is the importance accorded to deep strikes to disrupt Warsaw Pact reinforcement capabilities, with the long-standing operational objective of interdiction. The general premise of these deep-strike concepts is that, given sufficient time to mobilize, NATO forward-deployed forces can probably cope with the first operational echelon of a Warsaw Pact offensive, but that follow-on forces composed of subsequent eche-

83

lons, reserve forces and operational manoeuvre groups (OMG) would ultimately overwhelm the defences, penetrating deep into NATO's rear areas. This, coupled with massive air strikes against NATO airfields, air defence sites, command-and-control facilities, and nuclear weapons storage sites could preclude NATO reinforcement and force a collapse of Western defences.[3] This section describes the origins and components of three very distinct deep attack concepts which, because of common elements, have often been confused with each other in the recent debate over strategy. These three are the SHAPE FOFA operational sub-concept, the US Army's official operational doctrine (AirLand Battle or ALB), and the speculative Counter-Air '90 concept.

All of these deep attack concepts have been influenced by several factors: a recognition of potential vulnerabilities in Soviet *blitzkrieg* strategy and rigid echelonment of follow-on forces; discontent, particularly strong in the US Army, with what many military planners felt to be an unduly reactive defence; and an appreciation of the potential of emerging conventional weapons systems to acquire and destroy a broader range of Soviet targets with greater effectiveness.[4] Thus, in the late 1970s, a number of military planners and civilian analysts became convinced that, by adopting certain operational innovations and exploiting key technological advances, NATO could neutralize Warsaw Pact strategy. By striking deep, NATO could delay and disrupt the flow of quantitatively superior follow-on forces, thereby isolating forces in the first operational echelon and maintaining a combat ratio in engaged forces more favourable to NATO.

AirLand Battle (ALB)

Debate over these military and technological developments and over the appropriate lessons of the Vietnam and 1973 Middle East Wars led to the US Army's adoption of a new operational concept for corps-level and subordinate units worldwide, known as AirLand Battle (ALB). This new doctrine was codified in Army Field Manual 100-5 of August 1982.[5] ALB is a major shift towards a concept of manoeuvre warfare and away from the Army's previous Active Defense doctrine which envisaged fairly static operations supported by massive firepower to repulse attacks by attrition. Many Army commanders in Europe had come to doubt the viability of a doctrine that called for massing US firepower and manpower in a linear static fashion to cope with wave after wave of Warsaw Pact forces. The military thinking underlying the ALB concept reflects and can be traced to the work of the so-called 'defence reform movement'. Indeed, the reform movement, with its emphasis on mobility and manoeuvre and more effective exploitation of Warsaw Pact weaknesses and Western strengths, has had a major influence on several important developments in this area.[6]

Deep strikes within the corps commander's area of influence, extending to about 150 km ahead of the FLOT in support of manoeuvre operations, are only one facet of the ALB concept. ALB calls for a defence which secures and retains the initiative; breaks the momentum of the enemy's thrust by destroying synchronization among the elements of attacking forces; allows subordinate commands considerable flexibility within an overall battle plan; and utilizes the entire depth of the battlefield to strike an enemy and thereby prevent him from concentrating firepower or manoeuvring forces to points of his choice.[7] The two most controversial aspects of ALB are its concepts of conducting 'deep' strikes in order to delay, disrupt or eliminate uncommitted forces and to isolate committed forces, and of preparedness for the conduct of operations with chemical, nuclear and conventional weapons; that is, in a 'fully integrated' battlefield.[8] These two features, coupled with the manual's discussion of 'retrograde' operations, including delays where units trade space for time, triggered considerable alarm in Europe.

ALB requires close coordination between the conduct of the forward battle against committed forces and the 'deep battle' against follow-on forces. While ALB advocates a balance between firepower and manoeuvre, the latter is paramount because it enables the concentration of US strength against enemy vulnerabilities. Deep strikes are critical to the implementation of ALB in support of manoeuvre operations.

The ALB concept is not wholly dependent on the availability of the array of emerging technologies discussed above. Indeed, as one of ALB's architects notes, it was the need to integrate a variety of new weapons systems such as the M-1 tank, M-2 and M-3 fighting vehicles, and the AH-64 advanced attack helicopter into Army operations that provided the major impetus for the shift in doctrine.[9] The Army has also developed a speculative concept that explores how still emerging technology might affect the nature of conflict in the early years of the next century, called Army 21 (once called ALB

2000). Army 21 has often been confused with the ALB concept, and is the source of some of the wilder accusations about Army plans. In fact, the Army's near-term modernization, which calls for the purchase of 14 new weapons systems including such systems as the AH-64 helicopter and M-1 tank, has clear priority over the acquisition of emerging technologies.[10] Nevertheless, ALB operations, particularly the deep-attack missions, could be performed much more effectively if many of the aforementioned developments in target acquisition were available to the Army.

Many of the capabilities for target acquisition in ALB and most of the platforms for attacking deep to perform battlefield air interdiction (BAI) missions are provided by the Air Force. In addition to air support, however, artillery, missile strikes, deception, electronic warfare and manoeuvre of ground forces are all intended to play their part in ALB. Procedures for allocation of air support to ALB, particularly in NATO, are less than optimal. Allocation of air assets in Europe has been highly centralized, a practice that would restrict implementation of ALB because of the latter's requirement for the timely commitment of airpower. The US Army and Air Force reached an agreement in December 1982 on a document called 'Joint Air Attack of the Second Echelon'. This accord set up procedures to establish priorities among Army air interdiction requests and to provide the Air Force with a better appreciation of ground manoeuvre requirements.[11] This process was further advanced in 1984 when the Air Force formally endorsed the ALB doctrine. The two service chiefs, Generals Wickham and Gabriel, signed a memorandum of understanding that created a number of mechanisms to coordinate their respective operations and weapons development and procurement programmes in support of air-land combat.[12] This so-called 'Wickham–Gabriel Agreement' – and the 31 initiatives it endorsed – were heralded as the initial step in the development of a long-term process to guarantee the fielding of the 'most affordable and effective airland combat forces'.[13] The two services pledged that programmes supporting joint air–land combat operations will receive high priority in their respective development plans. They also agreed to exchange a formal priority list each year of those sister service programmes essential to their conduct of air–land operations, and to resolve joint or complementary system differences prior to programme development. This accord did not resolve all the differences the Army and the Air Force have with respect to air–land operations. The Air Force still does not endorse all aspects of the ALB concept, leaving some confusion as to how a war might be fought in the US corps sectors in Europe.

The SHAPE Follow-On Forces Sub-concept

In 1979, the then SACEUR, General Bernard W. Rogers, tasked the SHAPE staff to develop a plan for reducing to manageable proportions the size of Warsaw Pact forces that could reach NATO's general defence positions at the outset of any conflict.[14] Early studies at the SHAPE Technical Centre focused on improving air interdiction capabilities and attacking the second operational echelon of Warsaw Pact forces. Measures to enhance conventional capabilities were included in the 1981 ACE Force Proposals submitted to NATO's Military Committee.

Recognition of changes in Soviet doctrine, specifically the inclusion of highly mobile and potent OMG deployed within the first operational echelon, turned SHAPE's attention to targeting all troops from just behind the line of contact to as far into the Pact's rear areas as target acquisition and conventional weapons systems would allow. The FOFA concept envisages tracking Warsaw Pact forces during all stages of their deployment and targeting them at times and locations where they are most vulnerable. SHAPE planners recognized that several vulnerabilities in the Warsaw Pact's reinforcement process could be exploited to disrupt the momentum of an offensive. Because of the limited number of attack corridors, forces would be very densely concentrated and hence provide particularly lucrative targets in these areas. Similarly, along these attack corridors are a number of critical choke-points (rail-to-road transfer points, weapons depots, airfields, river crossings, and so on) where attacks would be particularly effective in slowing the forward movement of reinforcements and would create targets in the form of forces piling up behind these disrupted choke-points. The success of a Soviet advance would depend on the movement of various echelons to the line of contact on fairly rigid schedules. Delay, disruption or destruction of major elements of these follow-on forces could buy NATO the time it needs to reinforce its own front-line forces and ensure that the weight of the attack did not become overwhelming.

In developing the FOFA concept, the SHAPE staff reviewed operational trends in the forces of all NATO countries, including the US Army's ALB doctrine. The FOFA concept incorporates elements from the operational practices of all NATO countries and subordinate commands that are most applicable for ACE. FOFA is thus a multinational concept consistent with NATO's defensive posture. Indeed, as General Rogers has testified, some aspects of ALB, which was developed for global application, are not appropriate for ACE's defensive posture.[15] FOFA is designed only to improve ACE's capability to slow Warsaw Pact reinforcement capabilities throughout the theatre. The national corps of Allied Forces Central Europe (AFCENT) will still engage in combat according to their national operational and tactical plans.

FOFA is consistent with the NATO Forward Defence concept in that it seeks to stop reinforcing Warsaw Pact units *before* they reach NATO territory. Forward Defence does not limit itself to defending the small strip of territory at the inner-German border. As General Rogers noted:

> Our commitment to forward defense can best be implemented, in my view, once NATO is attacked, by carrying the battle forward – to the enemy's own territory. We continue, as you know, to evaluate improved methods for stopping and punishing an aggressor as deeply in his own territory as possible in order to preserve the credibility of forward defense.[16]

Thus criticism that FOFA is a new strategy is unjustified. NATO has always planned to interdict the forward movement of Pact forces. Moreover, this attack of follow-on forces early in the battle is designed to complement efforts to maintain NATO's general defence position. The key question is whether the high costs of procuring the systems necessary to pursue FOFA will divert critical resources from efforts to improve NATO's initial defences. Several studies have led the *Bundeswehr* to conclude that strikes beyond 150 km forward of the FLOT are not cost-effective.[17]

On 9 November 1984, the NATO DPC approved SACEUR's Long Term Planning Guideline for FOFA. With this authorization, various NATO bodies began deliberations over the means to implement this 'sub-concept' of ACE's overall concept of operations dictated by the strategy of Flexible Response. While NATO has always wished to attack follow-on forces, it has lacked the surveillance capabilities and weapons delivery systems – other than manned aircraft which would have to penetrate dense air defences – with sufficient range, accuracy and lethality to accomplish this mission. The FOFA concept applies to the entire ACE area and envisages strikes across corps boundaries, if necessary, up to approximately 300 km in front of the general defensive position. FOFA demands a highly centralized direction of all deep-strike capabilities to separate the aggressor's first and second echelon forces, thereby maintaining a manageable balance between NATO and Warsaw Pact forces in contact.

Warsaw Pact OMG and other options for employing follow-on forces are naturally high priority targets for FOFA. Considerable controversy persists about the OMG and their role in Soviet strategy.[18] However, SACEUR has determined that 'Much of the new target detection and sensing capability we seek to acquire is necessary for us to identify which forces are organized as OMG so they can be attacked early on.'[19] Indeed, it is fair to say that FOFA is somewhat more dependent on the development of a number of the emerging technologies described in Chapter 5 than is the US Army's ALB concept. As General Rogers testified in 1983, attacking follow-on forces 'requires the integration of intelligence collection, communication, data processing and dissemination to a degree that was not even conceivable a few years ago'.[20] Monitoring the extended battlefield with such systems as TR-1, JSTARS and US satellites and disseminating this information rapidly to operational commanders in a usable fashion are critical to FOFA's success. A number of the weapons systems essential to both FOFA and ALB, such as the MLRS, conventional *Lance*, TR-1 reconnaissance aircraft and a wide variety of attack aircraft, are already in the field.

While ALB and FOFA are generally compatible, there is none the less some competition for resources. While FOFA takes advantage of NATO's centralized mechanisms for allocation of airpower to achieve attrition in depth, ALB requires the rather early application of airpower in close support of ground manoeuvres. It is not clear whether such divergent requirements will force difficult trade-offs in allocating military resources. FOFA's deeper strikes, because of their delayed impact on the forward battle, may be judged less urgent than ALB's need to mesh air and ground operations against forces in contact. Very deep strikes against follow-on forces would clearly have a lower priority than dealing with a threatened breakthrough of NATO front lines.[21] Implementation of FOFA requires surveillance, target acquisition and attack systems capable of covering a much broader area than ALB, although some of the same collection platforms and weapons systems would serve both. ALB will require improved tactical C^3 and logistic support for manoeuvre warfare. Thus there is definitely some competition between ALB and FOFA in allocation of forces, and it would be well to ensure that systems designed for the deep attack of rear areas are still able to support the close-in battle if or when necessary.

Finally, it should be stressed that FOFA is not intended to replace NATO's options to employ nuclear weapons for deep interdiction or to alter the course of fighting elsewhere on the battlefield. FOFA is fully consistent with Flexible Response and its formal adoption in no way removes from the Soviet calculus the threat of nuclear escalation by NATO in response to any attack.

The differences between ALB and FOFA might seem to open the possibility of inconsistent operations by various national forces during a war. However, as General Rogers notes: 'Although some of NATO's national forces operate under tactical and operational doctrines and procedures which exhibit varying approaches to land combat, all forces which would come under SACEUR's command in the event of war would operate under ACE Chain of Command and ACE policies, doctrine, and concepts – not those of any single Alliance nation.'[22] He could hardly say any less. The fact remains that below corps level at least, there is no 'ACE doctrine', only a collection of national doctrines dictated by national traditions, experience and preferences. This has never seemed to matter unduly so long as the war is fought by national corps. However, with FOFA controlled at a higher level than corps, a greater degree of doctrinal convergence may well be required.

Very Deep Strikes and Counter-Air '90

In the early 1980s, the Director of Defense Research and Engineering (DDR&E) in the Pentagon developed a concept to improve the suppression of Warsaw Pact air power called 'Counter Air 90'. Among other things, Counter-Air '90 envisages very deep strikes with stand-off weapons or ground-launched ballistic missiles, at ranges even beyond 300 km, against both mobile and fixed targets of relevance to the war in the air. The ESECS study cited earlier advocates some elements of this concept, such as airfield attack. During 1983, the Principal Deputy Under-Secretary for Research and Engineering, James P. Wade, briefed this proposal to a number of European officials.[23] Some of these same officials, who had seen presentations on ALB, ALB 2000 (Army 21), and the Rogers FOFA plan, were either very confused or viewed all these concepts as part of a comprehensive package. This interpretation justifiably caused considerable anxiety among both officials and the attentive public.

Counter-Air '90 is a total systems-engineering approach designed to utilize offensive and defensive weapons in an integrated air and missile defence system. It includes a scheme for networking various radars, attacking Warsaw Pact electronic warfare (EW) aircraft with longer-range missiles, developing a new IFF system and improving C^3.

The proposal also includes the development of a two-tiered anti-tactical missile (ATM) system to deal with the growing Soviet capability to attack NATO's *Patriot* and *Hawk* air defence systems with ballistic missiles. The first element of this system would be a weapon, such as the upgraded *Patriot* system described in Chapter 5, to intercept incoming tactical ballistic missiles. US analysts are also looking at the possibility of using non-nuclear kill mechanisms developed for terminal defence against strategic ballistic missiles for this mission. The second aspect of this ATM programme involves developing surveillance and targeting capabilities to destroy Warsaw Pact tactical ballistic missile *launchers* before they are able to reload. Long-range conventionally-armed missiles, provided with precision targeting information by a more capable version of the AN/TPQ-37 *Firefinder* weapons-locating radar, could be used to target missile launchers, missile stocks, and artillery.[24] A system mating a European-designed re-entry vehicle and warhead with the *Pershing* II or *Trident* missile's first stage has been suggested, but pursuit of these concepts has almost certainly been foreclosed by the INF agreement that eliminated long-range INF missiles.

A central feature of Counter-Air '90 is the notion elaborated in Chapter 5 of employing conventionally-armed long-range ballistic or cruise missiles armed with ICM to attack Warsaw Pact airfields at the early stages of a war.[25] The objective of these attacks would be to destroy Warsaw Pact capabilities to generate sorties and force aircraft recovery operations into areas more vulnerable to attack. While the US Air Force reportedly prefers to consider using cruise missiles for this role, most Department of Defense officials favour ballistic missiles because of their greater payload and range capabilities. Counter-Air '90 therefore involves not only the development of advanced passive and active defences of airfields and air defence assets, and of new C^3I and electronic warfare capabilities to achieve dramatic improvements in NATO's capabilities to suppress Warsaw Pact airpower, but also the attack of air bases deep inside Eastern

Europe. Doubts have, however, surfaced about the cost-effectiveness of Counter-Air '90 and its impact on intra-war and pre-war stability.

Concerns about Stability

Europeans and some Americans have expressed three principal objections to deep strike concepts. First, as noted above, a number of military analysts believe that deep strikes beyond 150 km are not cost-effective because the expensive surveillance and target-acquisition systems that they require can be easily disrupted or overcome. Second, many military analysts believe that the deep battle is not nearly so critical for NATO as dealing with the first echelon of a Warsaw Pact attack. Thus, it is argued, development of deep strike capabilities would risk directing funds from more pressing military requirements in the close-in battle. Finally, the conduct of extensive deep strikes has an offensive character that many feel is inappropriate for a defensive alliance. Some West Europeans worry that development of deep strike capabilities would raise fears in the East that NATO might seize Warsaw Pact territory, as has been advocated by several strategists. Such an operational concept, appearing to embody a more offensive orientation, is seen as potentially destabilizing. West European officials fear that actual or apparent acquiescence in the ALB or any retaliatory offensive concept would complicate their peacetime relations with the East, and hasten Soviet nuclear employment in wartime.

The fact is, however, that NATO has always planned for a limited number of deep strikes against fixed targets like airfields and key choke-points where there is a high probability of success. What has changed is that, with the advent of new types of accurate ballistic missiles, it now has at hand the technology to achieve this more effectively and against a wider range of targets. It is also important to consider these options in the light of strategic goals. For those who anticipate or advocate a short conventional war, deep strikes would probably not make a significant impact on the outcome, but they might, however, be pivotal in a long war. Similarly, NATO might not have to deploy extensive numbers of deep strike or other ET systems to achieve the military benefits of forcing the USSR to alter its operational concepts and tactics, or implement costly counter-measures throughout Warsaw Pact forces, or both. For example, NATO possession of some top attack capability against tanks might force the Warsaw Pact to take defensive measures throughout their entire force in an armoured assault because of uncertainty as to which segment of their force might be attacked with this weapon system. Several other implications of these new military concepts have yet to be considered adequately. For example, it is often suggested that so long as NATO maintains dual-capable systems, deep strikes with conventionally-armed ballistic or cruise missiles may present escalation risks as great as the conduct of such operations with nuclear systems. The sight of 2000 ballistic missiles climbing over the Western horizon, it is argued, would hardly foster calm responses from Warsaw Pact headquarters. Despite the apparent cogency of the argument, this may not necessarily be the case. If the USSR hopes to avoid nuclear escalation if at all possible in a war in Europe, as is suggested in Chapter 8, they might very well want to ride out a large ballistic missile attack. In fact, the very size of such a large attack, even with dual-capable missile systems, might be a fairly good indicator of its non-nuclear character. It may matter much more what the target of any such attack is than what weapon system is used. Extensive or even selective conventional strikes deep into Eastern Europe could reach a point where they threaten not only Warsaw Pact reinforcement capabilities but also military and governmental instruments of control. Faced with such a strategic threat, the Soviets would probably find retaliatory or even pre-emptive strikes with nuclear weapons attractive. Secure control of their East European buffer zone is clearly a vital Soviet national interest, which would be defended with all means available.

This escalation risk also raises the question of political control. Would SACEUR have independent authority to launch such strikes or would he have to consult with Allied Governments to obtain authorization? As early as 1983, the governments of several European NATO states declared their opposition to the development of long-range conventionally-armed cruise missiles on their soil, citing concerns with these escalation risks and the serious problems that proliferation of such missiles could present for nuclear arms control in the region.[26] It is possible to foresee considerable domestic political opposition, on the scale of the anti-INF crusade, to the construction of bases for 7000–10 000 conventional attack missiles. All these questions will have to be scrutinized with care as NATO decides on these particular conventional defence options.

RATIONALIZATION OF MILITARY MISSIONS AND MORE EFFECTIVE USE OF RESERVES

Another general category of conventional force improvement proposals involves changes in operational concepts and improved defence cooperation among NATO states. The concepts range from fundamental restructuring of various national force components in order to achieve the most efficient use of resources in the conduct of specific military missions, to more effective use of the very substantial pool of trained manpower that exists in the form of European reservists. Indeed, some would argue that the *only* way that NATO can expand its conventional forces is through such restructuring. There is no reason to take for granted the fact that NATO spends more and yet obtains less conventional defence than the Warsaw Pact.[27]

A number of proposals have circulated in recent years calling for a greater rationalization of military missions assigned to various NATO member states. There is little question that some military missions are being conducted inefficiently or ineffectively by states ill-suited and ill-equipped to fulfil them. In other cases, reallocation of scarce resources to areas of comparative advantage from areas where a state's military infrastructure and capabilities are weak could yield more defence at existing levels of expenditure, For example, some analysts have suggested that the West German *Bundesmarine*, which presently performs various missions in both the North and Baltic seas, could scale back to Baltic deployments only and transfer its North Sea responsibilities to the British and Dutch navies.[28] The resources saved by this could be allocated to improving various *Bundeswehr* capabilities. Others have argued that the missions of the surface vessels of the smaller navies of NATO – the Belgian, Danish, Dutch and West German – could easily be accomplished by ships from the US and British fleets. However, as Ingemar Dorfer has noted, the savings from the elimination of these four small surface fleets would be marginal.[29] Moreover, the Dutch navy is of high quality, and the Danish and German surface vessels operating in the Baltic have important roles in war and peace.

It might be more valuable and politically feasible to restructure European air forces and allow the US Air Force to assume a greater share of the burden for NATO tactical air power. Modernization of West European air forces over the coming decade will be very costly. The West German Air Force will need to replace about 200 aircraft.

More than 340 F-16 fighters are entering service with the Belgian, Dutch, Danish and Norwegian air forces, and additional aircraft are on order. This is taking place despite the fact that only 1500 of the 5500 US combat aircraft available would deploy to Europe in wartime to augment the force of 850 stationed there in peacetime.[30] Dorfer argues that if the US were to base only six additional wings (432 aircraft) in Europe in peacetime these four Allies could gradually reduce the size of their air forces through natural attrition and still retain considerable fighter forces. Dorfer contends that this forward basing would not cost any more than operating these aircraft from the US, and that the savings realized by the four European governments could be used to create an additional twelve West German, three Dutch, and two Belgian divisions for the Central Region.

In addition, it is clear that NATO has a very large pool of trained manpower for which no explicit wartime roles are envisaged. The critical questions concern how these resources can best be organized and mobilized and at what cost. Steven Canby and Ingemar Dorfer have estimated that, for a 20 per cent increase in its defence budget, or redirected savings of the same amount, the Dutch Government could establish two additional active armoured brigades and base another forward in West Germany.[31] These forces could be incorporated into the efficient Dutch RIM reserve system. The RIM programme, whereby the same cadre of professional personnel stays with the unit in both its active-duty and its reserve phase, can provide four more brigades within 36 hours of mobilization. Under this scheme, the Dutch could then provide a total of seven tank, six mechanized and six motorized infantry brigades, substantially strengthening their corps sector in NORTHAG. If this concept were implemented in tandem with some restructuring of NATO forces, it is argued, the 30 per cent savings realized from the latter would more than cover the cost of the former.

In a similar fashion, Canby contends that the West Germans could easily add 12 brigades, creating a force of six corps of 12 mechanized and 12 motorized infantry divisions by more effective use of reserves at an additional cost of about 15 per cent of current defence expenditures. This increase in costs, it is estimated, could also be covered from savings arising from reduced air force procurement, as could the creation of two new Belgian divisions. Such a force, the two analysts argue, would help to deter attack by providing the in-place battlefield reserves that NATO would need to back up front-line manoeuvre forces and to buy time for US-based reinforcements to arrive in Europe.

Canby has advocated utilizing some of these expanded reserves to form light territorial units which could be effective in defending the two-thirds of NATO's forward area which consist of close terrain, thereby freeing armoured forces for concentration on the open areas of the North German plain and in operational reserves. These territorial units could engage in static defence of the many villages that lie athwart likely Warsaw Pact axes of advance and could also conduct mobile operations in the many heavily wooded areas.[32] An alternative approach of this kind would mean that territorial units would occupy and control the forests and villages while active duty *Jaeger* brigades limited movement through these areas. Such operations could be used to channel enemy movement, attack mechanized forces and mask reserves.

One final restructuring measure advocated by Canby and others concerns increasing the ratio of combat to support troops in NATO divisions. While a number of NATO armies have increased their combat strength, all NATO countries still have larger 'tail-to-teeth' ratios than comparable Soviet formations. By slimming its divisions and 'division-slices' (the associated support and logistic elements for the divisional organization) down to 20 000 troops, Canby estimates that NATO could (counting French forces) field 75 instead of the present 33 divisions on the continent. These units would, of course, have to be equipped, but Canby contends that this would not be a major expense since some of this could come from stockpiles.[33] There would, however, clearly be penalties in sustainability and the replacement of battle losses if such war maintenance reserve stocks were not made good.

Achievement of such sweeping alterations of traditional national military missions and operations faces formidable practical and political hurdles. Nevertheless the current political climate, with growing European defence cooperation and scarcer resources, may be conducive to the realization of such of these initiatives as prove genuinely feasible. It is not clear how effective reserve units could be in substituting for regular forces without much higher levels of refresher training and readiness than currently exist in most NATO countries. Moreover, making most NATO reserve units more combat-ready would require substantial new investment.

7 A Strategic Concept for Conventional Defence

THE SEARCH FOR A COMMON STRATEGY

NATO's consideration of its conventional defence options has been bedevilled by the absence of a clearly specified strategic concept commanding unanimous support. Members of the Alliance have differing views of the requirements of deterrence and of the most appropriate military objective for its forces if deterrence fails. Differences among the Allies about the operational implications of the doctrine of Flexible Response were swept under the carpet by the ambiguous language of Military Committee (MC) Document MC 14/3. Since the adoption of that doctrine, most Allied leaders have assumed that it would be too costly politically, and probably undesirable militarily, to achieve a more specific articulation of NATO's strategic concept. As a consequence, defence planners make decisions about conventional options and priorities on narrow cost-effectiveness grounds guided by varying interpretations of this imprecise strategic concept.

There are a number of strategic possibilities within and beyond the context of Flexible Response, ranging from a tripwire posture to one of indefinite conventional defence, which should be considered in the light of the various constraints and opportunities which exist. Such a review should facilitate consideration of whether various proposed conventional defence options are not only practicable but also appropriate to desired military objectives.

This chapter considers five strategic concepts which could guide NATO conventional defence planning: a very brief conventional war plan envisaging the threat of early nuclear use; SACEUR's proposals to extend the non-nuclear phase of any conflict by developing a more robust conventional posture designed to be sustained for 2-3 weeks before turning to nuclear weapons; development of new capabilities to sustain a conventional war for at least 40–60 days – as the US Defense Guidance documents have reportedly called for – or perhaps even indefinitely, or until the Warsaw Pact employs nuclear weapons; adoption of an offensive retaliatory strategy, as advocated by Samuel Huntington; and alteration of military capabilities in ways that minimize offensive potential, the so-called defensive defence posture advocated by some left-wing political parties, notably in West Germany. The first three concepts are consistent with Flexible Response: the latter two would require changes in Alliance strategy. NATO's lack of an unambiguous and unanimously endorsed strategic concept has hardly gone unnoticed. The broad array of proposals for improving the Alliance's conventional posture advanced in recent years has created a good deal of confusion among both Western governments and publics. This situation led West German Defence Minister Manfred Woerner and other European officials to lament the absence of a conceptual framework to provide Allied planners with guidance on goals. It was in response to these concerns that NATO's DPC adopted in May 1985 the short-term action plan for CDI discussed earlier, which identifies priorities for enhancing general purpose forces to encourage convergence between Alliance and national defence planning;[1] and subsequently, in December 1985, the defence ministers endorsed the CMF, setting forth ACE military plans for the next 20 years, and designed to provide nations with 'broad, longer term guidance on the military requirements of NATO strategy'.

CDI and CMF may ultimately prove helpful in sorting out some of the new conventional defence proposals. However, they are essentially only resource allocation plans designed to rectify broadly defined shortcomings in NATO's conventional posture within prevailing national budgetary constraints and vague articulations of Alliance strategy. The CMF endorses the SACEUR's principal plans for enhancing NATO's conventional capabilities in ways that reduce the likelihood of early first-use of nuclear weapons. However, it fails to articulate the strategic objectives of conventional defence and the military priorities that follow therefrom.

In contrast to this conservative approach, which does nothing to resolve ambiguous and debatable strategic assumptions, this chapter considers the fundamental requirements of conventional deterrence and Allied military war-fighting strategy in the event that deterrence should fail. Is it necessary for deterrence that NATO's conventional forces should palpably be capable of defeating Warsaw Pact forces? Or would it be sufficient for NATO forces to have only the capability to deny the Warsaw Pact a quick, low-cost military victory, the prospect of which most analysts see as the most likely precondition for such an attack?[2] Once the decision has been taken as to the strategic goals of conventional defence, choices on operational issues, such as deep strikes and new weapons systems, can be made much more prudently and with a better understanding of the contribution they can make towards the overall objective.

Answers to these questions on strategic goals depend somewhat on views of how deterrence might fail. Twentieth-century military history suggests that calculated aggression is likely to be deterred if the attacker believes that the defender can wage a protracted war of attrition. If it is assumed that the Warsaw Pact might attempt to exploit Western weakness to achieve a limited military objective, then denial of an expedient victory would seem sufficient. If it is the view that war in Europe would only start accidentally, the threat of escalation rather than necessarily that of military defeat may be the more compelling force for rapid cessation of hostilities.

NATO's strategic choices must be examined against the background of Warsaw Pact options. One such option that the Pact has is to maximize surprise by initiating a standing-start, unreinforced attack against NATO using the 19 Category 1 divisions in the Group of Soviet Forces Germany (GSFG) to seize, for whatever reason, perhaps 100 km of West Germany. The more likely threat to NATO would be posed by a Warsaw Pact offensive that followed a 14–28 day mobilization period which generated as many as 100 divisions. In conjunction with either of these moves, the Pact would use its air and naval power to disrupt NATO's reinforcement potential. Each of these Eastern options creates different challenges for NATO strategy.

STRATEGIC OPTIONS WITHIN FLEXIBLE RESPONSE

NATO's doctrine of Flexible Response establishes several minimum requirements for conventional forces. First, they must deter aggression by undermining the Warsaw Pact's confidence in achieving a strategic success by conventional attack. NATO's conventional defences must be sufficient to deny Pact planners a reasonable prospect of rapid success. Second, should deterrence fail, they must be capable of defending forward against an unreinforced conventional attack without the use of nuclear weapons, and of restoring the status quo ante.[3] Finally, they must be able to sustain a forward defence that will disrupt a reinforced attack and cause the termination of aggression and withdrawal from NATO territory, or afford NATO the time to contemplate (and possibly execute) deliberate escalation and whatever other actions might be necessary to terminate hostilities and force a withdrawal. NATO planners are convinced that existing Allied conventional forces could never achieve the capability to defeat consecutive attacks by undiminished Warsaw Pact first and follow-on echelons. Within the broad concept, the exact required or planned duration of a conventional defence should not be stated precisely in order to preserve the incalculability that is a vital element of this strategy. None the less, NATO planners argue that the longer a purely conventional defence can be waged the greater the range of options open to the Alliance.

A Conventional Tripwire

One strategic concept which could guide NATO's defence choices would be to maximize reliance on nuclear deterrence by planning for a very brief conventional response to any aggression before threatening nuclear escalation. The notion that NATO plans call for a brief conventional 'pause' has always been the preferred interpretation of MC 14/3 for many, particularly in Europe, who find the nuclear spectre the most effective deterrent of any aggression. Most European strategists abhor the notion of a protracted conventional war given geographic realities and the enormous destructive power of modern non-nuclear weaponry. NATO has little depth for the conduct of a long war, and the density of development in Central Europe means that *any* war would result in societal devastation.

Many NATO officials contend that this tripwire posture has become, *de facto*, Alliance strategy. As former SACEUR, General Bernard W. Rogers, stated on many occasions: 'NATO's current conventional posture does not provide our nations with adequate deterrence of Warsaw Pact non-nuclear aggression or intimidation derived from the threat of such aggression. If attacked conventionally today, NATO would face fairly quickly the decision of escalating to a nuclear response.'[4] General Rogers has noted that NATO finds itself in this situation for a number of reasons, primarily because of its inability to sustain Allied forces adequately with trained manpower, ammunition, and war reserve materials. Rogers and many other NATO officials have argued that this extreme reliance on the nuclear response strains the credibility of Western deterrence in an age of superpower nuclear parity. The SACEUR would prefer to have a 'reasonable prospect' of repulsing a non-nuclear attack by conventional means. The possibility of nuclear retaliation would remain, as would some uncertainty about its timing, as an ultimate guarantor of Western security.

However, others argue that the possibility of rapid escalation is an effective deterrent to any aggression. Advocates of this posture contend that raising the nuclear threshold, particularly to some specific demarcation point, actually undermines deterrence by making limited acts of aggression less risky. Moreover, it is argued, such a posture provides the tightest link to the American strategic umbrella, which is the foundation of Western deterrence of Soviet aggression. Many Europeans, on the basis of rational calculations of national self-interest, may doubt the credibility of the American nuclear guarantee in the age of superpower parity. Nevertheless, they still believe that as long as American strategic and theatre nuclear weapons remain a factor in the security calculus, the Soviet Union will be deterred from aggression in Europe because Soviet leaders cannot be *certain* that there will not be nuclear escalation in a conventional war.

As noted, French strategic planning has been guided by a concept of fairly rapid escalation to 'pre-strategic' nuclear warning strikes, with conventional forces relegated to an ancillary role. French forces are prepared for relatively short periods of combat to counter minor aggressions and to assert the national will to resist. The French First Army lacks the conventional battlefield sustainability of the US Seventh Army and the *Bundeswehr*, and many French strategists believe that efforts to enhance the Army's strength actually undermine deterrence.[5]

Advocates of this concept favour at most only *marginal* improvements to NATO's general purpose forces, designed to preclude automatic escalation while maintaining the coupling of European security to the US nuclear guarantee. NATO's military priorities under such a posture are the maintenance of strong front line conventional forces along the main axes of likely Warsaw Pact attack and the protection of theatre nuclear assets to ensure that these systems can accomplish their strategic warning or battlefield interdiction missions. Conventional deep interdiction strikes against follow-on forces and counter-air operations would be of marginal value in supporting this posture.

The risks of rapid and ultimately uncontrollable nuclear escalation inherent in a tripwire posture necessarily mean that this concept suffers from severe credibility problems if deterrence fails. It is also difficult to comprehend how conventional military forces can be motivated to fight if they lack a strategic objective other than triggering this self-defeating escalation. Moreover the concept affords precious little time in a crisis for political authorities to negotiate some resolution. For all these reasons, it is unlikely that Western military and political officials would have the will actually to execute such a strategy. If the Soviets operated on the basis of a similar assessment, deterrence would be undermined. None the less there remains some elite support, particularly in Europe, for this concept.

No Early First-Use

General Rogers and a number of prominent European strategists have suggested that incremental improvements to NATO's current posture would enable it to sustain conventional deterrence for an unspecified period and avoid early first-use of nuclear weapons. Such a posture would not eschew first-use of nuclear weapons and would therefore retain the close coupling of European security to the American strategic guarantee. However, maintaining the credibility of this posture requires that the Alliance have a 'reasonable prospect' of frustrating a Warsaw Pact conventional attack by non-nuclear means. This concept has achieved considerable attention because it has broad elite support and would appear to be relatively affordable given current resource constraints. Moreover, General Rogers and others have set out fairly specific force enhancement priorities designed to enable NATO to sustain such a concept.

The foremost priorities in this context would be to enhance NATO's sustainability by procuring more equipment and by improving the maintenance, training and support for existing forces committed to ACE.[6] These forces will occupy NATO's forward defensive positions and must be equipped to cope with the leading echelon of a Warsaw Pact attack. The second priority in this regard would be modernization of NATO's weapons systems, particularly those for attack of follow-on forces, so as to delay, disrupt and reduce to a manageable ratio the subsequent echelons of Warsaw Pact troops reaching NATO's forward defence positions. The third priority in this scheme is to augment NATO's force structure which, given prevailing demographic trends, would require the integration of better trained ready reserves into NATO planning.

West Germany's current defence plans have similar goals, but with slightly different priorities. The *Bundeswehr's* first priority is to improve its ability to conduct initial defence against a short warning attack as close to the frontier as possible.[7] German planners see improved NATO air defences and more effective offensive counter-air operations as the second most pressing requirement. As a third area for attention the *Bundeswehr* will seek to improve its capabilities against follow-on forces in the Warsaw Pact's first strategic echelon. Finally, the German plan seeks to enhance the sustainability of Forward Defence against the second strategic echelon by developing capabilities for combat in depth to attack these forces in the deployment phase.

Thus maintaining a more prolonged non-nuclear defence of Central Europe would require NATO forces to be able to deal with both short-warning, unreinforced attacks, and the fully mobilized Warsaw Pact threat judged to be the more probable contingency. In this latter scenario, William Kaufmann has estimated that within 14 days of mobilization and deployment the Warsaw Pact could field as many as 90 combat-ready divisions across Central Europe.[8] Enhancement of Allied all-weather, day and night surveillance and target acquisition systems, over both short and long ranges, would be vital to the success of this defence concept. Operational reserves would have to be expanded to deal with Pact breakthroughs and the operations of special units in rear areas. Such a concept would also require fairly extensive protection of nuclear assets, airfields, air defences and logistic facilities. NATO's ability to deal with threats to its own reinforcement potential and C^3I assets as well as disrupting corresponding Eastern targets would also be an important consideration. NATO's position in this situation would improve markedly if France made a clear commitment of a substantial force.

The SACEUR's FOFA sub-concept seeks to attack Warsaw Pact forces from just behind the troops in contact to as far into the enemy's rear as NATO's target acquisition and conventional weapons systems can reach. Some of this mission can be performed with existing missiles and aircraft such as *Lance*, MLRS, *Tornado* and the TR-1 reconnaissance aircraft. However, striking into the second strategic echelon with a high degree of success will require the acquisition of new ballistic and cruise missiles and improved target acquisition systems. The effectiveness of FOFA will be influenced by the degree of NATO's success in other missions, such as establishing air superiority and suppression of air defences. Conversely, however, in circumstances of scarce resources, the value of FOFA – even within the context of moving towards a posture of no-early-use of nuclear weapons – has to be weighed against the merits of increased investment in other missions which might also buy the Alliance more time before escalating. Consideration of this trade-off was a central factor in the *Bundeswehr's* decision to make the improvement of their capability to wage the close battle and air superiority missions their top two priorities.[9]

The most difficult operational question associated with this concept concerns the development of criteria for escalation or the threat of escalation. Clearly there can be no absolute conditions as to how long conventional defence should be waged. None the less some general assumptions about this threshold must be made to guide planning for such things as the size of war reserve stocks and reinforcement efforts. Escalation decisions will ultimately be made more on political than military grounds. Because the strategic objective of this concept is to counter aggression in a proportionate fashion that allows time for a negotiated settlement, it will be the political authorities who must decide when the utility of the conventional response is exhausted and escalation is required.

Protracted or Indefinite Conventional Defence

The most ambitious conventional defence concept consonant with Flexible Response would be to develop the capabilities to sustain a non-nuclear conflict for 40–60 days, as reportedly called for in the US Defense Guidance, or until the Soviets use nuclear weapons. Indeed, American military planners often assert that because it is possible neither NATO nor the Warsaw Pact will use nuclear weapons, plans must be developed for a conventional war of indefinite duration. A policy of no-first-use of nuclear weapons would require a similar posture. Executing such plans would require not only a sustained build-up of NATO conventional capabilities in the Central Region, but also effective protection of ports and sea lanes. In addition to the initiatives mentioned with respect to no early use, it would probably require further augmentation of forces in forward areas, more effective mobilization, outfitting and replenishment of reserve forces, expanded sealift, protection of primary and secondary reinforcement potential, particularly sea lanes and ports, and a wide array of other measures. Thus restoration of the NATO LOC through France, as a hedge against the loss of Antwerp and Bremerhaven, might be particularly valuable in a long war.

It is possible to question why it might be seen as being preferable to plan for such a conventional war in Europe, which would inevitably wreak mass destruction of society, given that even after such a cataclysmic struggle the war could still become nuclear? Warsaw Pact forces with their large equipment stockpiles are configured for a long war, but the Eastern industrial base would be at a severe disadvantage in dealing with fully mobilized Western economies. [10] Moreover, if engaged in such a long war, there could be a growing possibility that Moscow would have to cope with societal upheaval in Eastern Europe and increased vulnerability to Chinese military action. As a result, there might be greater pressures on the Soviets in a long war to escalate and achieve decisive victory.

Nevertheless, with a longer period of mobilization Warsaw Pact forces could pose an even greater threat to NATO than in the case of shorter warning attacks. For example, one estimate is that 120 days after mobilization the Soviets could field 110 divisions as against 46 for NATO. [11] Force-to-space constraints might restrict Soviet ability to bring all this manpower to bear simultaneously on the tactical situation but at the very least it would clearly enhance Warsaw Pact scope for maintaining the impetus of their offensive by the successive commitment of fresh echelons to replace spent forces at the point of contact. In order to rebuff such a threat, NATO would need to augment its manpower by substantially expanding the ready reserves. It is unclear whether NATO could find the trained manpower to replace combat casualties and allow for rotation of battle-weary units in a conflict that lasted more than a month. [12] The US would probably have to institute conscription to meet the demands of such a posture.

Preparing for such an indefinite conventional defence would require maintenance of greatly increased war reserve stocks of munitions and other military supplies in order to sustain a defence until the civilian production base could be put on a war footing. The US is currently pursuing a programme to provide 60 days' worth of war reserve stocks. However, other Allies are disinclined to follow suit, and would have to acquire the requisite supplies. It is generally assumed that such an indefinite defence would require stocks sufficient for 180 days in order to sustain Allied forces until production lines could reach a war footing and keep pace with combat consumption. [13]

There are several other options that NATO might pursue to implement this long war concept. In most force comparisons, NATO does much better against all likely Warsaw Pact threats by making greater use of barriers and terrain enhancement measures. [14] Effective barriers would slow the advance rates of attacking forces and could permit NATO to have a somewhat thinner defence at the front with more forces held in reserve to cope with breakthroughs. However, there has long been intense domestic resistance in West Germany to the erection in peacetime of any permanent barriers on the Western side of the inner German border, and the ability to construct any such obstacles during a transition-to-war phase would be heavily dependent upon the availability of time and resources. Within these political constraints, however, further preparations such as forward positioning of barrier materials, pre-chambering and limited terrain enhancements would be helpful even in short conflicts.

Improving NATO's capabilities to conduct conventional deep strikes against follow-on forces and air assets would be essential to successful implementation of this concept. Suppression of Pact aircraft and air defences could clearly have an important impact on the course of a long war. Similarly, the delay, disruption and attrition of follow-on echelons – even those fairly deep in Warsaw Pact territory – with aircraft or missile delivered ordnance could have a dramatic impact on force ratios at the front. For example, such deep strike missions would restrict the movement of reinforcements and replacements, force the dispersion of defences and require units to be held in the rear to repair damaged lines of communication. An optimal campaign of air superiority and interdiction might actually destroy the Pact's capability for offensive operations after several weeks. Similarly, shutting down some or all of the Pact's MOB and DOB would seriously degrade the offensive air threat to NATO, which would be particularly important in a longer war.

COUNTER-OFFENSIVE OPERATIONS

In the midst of the controversy over various official concepts for the conduct of deep strikes, Professor Samuel Huntington advanced a proposal that NATO should adopt a conventional retaliation strategy.[15] Huntington proceeded from the premise that a credible deterrent must threaten an aggressor with punishment as well as denying him his objective. Before the advent of nuclear parity between the superpowers and Soviet conventional superiority in certain areas, NATO could, he has argued, credibly threaten such retribution with nuclear weapons. Huntington believes that even if a major increase in NATO's conventional capabilities could be realized, this posture under the current strategy of direct defence – deterrence by denial – could not substitute for the deterrent effect of the threat of nuclear retaliation. Conventional defence improvements, he notes, do not necessarily enhance deterrence: 'To be effective, deterrence has to move beyond the *possibility* of defense and include the *probability* of retaliation.'[16] Thus, he concludes, NATO needs to develop a conventional offensive capability that could place in jeopardy Soviet control over Eastern Europe, which Moscow perceives as vital to its own security. Such a deterrent posture would be more effective than a denial strategy, where the costs of aggression are more calculable and subject to management by the attacker, because it would make the costs incalculable and largely beyond the aggressor's control.

Huntington asserts that a counter-offensive strategy would not conflict with NATO's politically defensive orientation. A defensive alliance, he argues, need not have a defensive strategy. NATO had precisely such an offensive strategy with plans for the use of nuclear weapons under the doctrine of massive retaliation. This argument does not address the impact of such a doctrine on stability. It is not only the West that infers political intentions from military capabilities. However, Huntington adds a measure of reassurance by suggesting that the threat of retaliation would be qualified and designed to exploit rifts between the Soviet Union and its allies. For example, NATO could assure East European governments that they would not be invaded if they did not cooperate with a Soviet attack in the West. Such a strategy would, he contends, also exploit the apparent inflexibility of Soviet war plans. Their force structure is less well equipped to defend and NATO could exploit surprise.

Finally, Huntington argues that the military requirements of such a strategy are not out of NATO's reach. He notes that despite a slight Soviet edge in numbers, NATO could employ mobility to mass forces at unexpected points. Thus he sees conventional retaliatory operations as entirely compatible with, and complementary to, the kind of deep strikes called for under the FOFA and ALB concepts. Similarly, Huntington claims that NATO's existing heavily armoured formations, currently justified for mobile defence and the conduct of counter-attacks, would be well suited for a retaliatory offensive.[17]

A study using dynamic war games roundly disputes this assertion about force requirements. Modelling suggests that execution of Huntington's strategy of horizontal escalation would require at least a doubling of American ground forces based in Europe and an enormous expansion of NATO's reinforcement potential.[18] Huntington has stipulated that the forces used in a counter-offensive would have to be in place and prepared to fight on the day war began. Thus it would be preferable for any American divisions to be based on the Continent.

Since reinforcements are an important element of current military plans, they would be even more vital if NATO were attempting to move large units into Eastern Europe while defending a large sector of its own territory. In fact, one danger inherent in this strategy is that some of the manoeuvre forces currently programmed to thwart deep penetrations of NATO forward defences would be diverted to the retaliation mission. The conduct of such deep operations would place extraordinary demands on strategic lift and tactical air support. Most importantly, this manoeuvre force would face a double challenge, as Keith Dunn and William Staudenmaier explain:

> Assuming a best case for the NATO force . . . it would require at least ten to fifteen additional US divisions to carry out this operational plan. Not only would the force have to maneuver and fight its way into Eastern Europe, but also it would have to defend itself from the inevitable counterattack of the Soviet second strategic echelon. In addition, it would have to defend a long and vulnerable supply line through hostile territory as well as a 360 degree defense perimeter. Each of these tasks is a difficult undertaking, particularly when the risk of nuclear war would be high. Trying to do all of them simultaneously is virtually impossible under current circumstances.[19]

Manning even four additional American divisions would probably require reinstitution of a peacetime draft, and the five-year cost of establishing and maintaining such a force in the US would exceed $40 billion.[20] Thus this strategy would require extraordinary military forces which are clearly beyond NATO's reach in the current environment of dwindling manpower and scarce resources.

If NATO were able to achieve military force levels sufficient to conduct counter-offensive operations, it would probably not want or need to alter its current strategy. A military capability of this magnitude would certainly shore up the non-nuclear component of Western deterrence. Huntington asserts that his proposals would salvage rather than replace current strategy. Noting the declining credibility of the nuclear component of Allied doctrine, he argues that the development of a conventional retaliatory capability would actually bolster flexible response by providing NATO with an option other than conventional defence in response to aggression. Similarly, he contends that this option would complement forward defence by moving some of the battle from West Germany to Warsaw Pact territory. In essence, Huntington argues, his proposal 'would in effect make flexible response more flexible and forward defense more forward'.[21]

The counter-offensive strategy is incompatible with current NATO doctrine for several reasons. NATO plans for Europe have the limited military objective of maintaining the territorial integrity of the Alliance by repulsing any Warsaw Pact aggression and securing the status quo ante. A counter-offensive strategy seeks a much broader goal: destabilization of Soviet hegemony in Eastern Europe. This strategy increases the risks of nuclear escalation by broadening the conflict and intensifying the nature of the superpower confrontation. If the Warsaw Pact moved on Frankfurt, it is unclear what effect seizing cities in Eastern Europe would have in achieving NATO's objective of restoring West German control of its territory. Such punishment might deter further aggression, but it would probably not force a Soviet withdrawal from West Germany; indeed, conflict termination might be made more difficult. The Soviets might have the upper hand in negotiations because their strategic depth would allow them to isolate the NATO forces in Eastern Europe more easily.

Finally, the European Allies abhor this strategy and the military posture it implies. Its offensive character, directed at their East European neighbours, is incompatible with the European view of NATO as a defensive alliance. Europeans fear such a strategy would be inimical to *détente* because the Soviets would interpret it as a threat to change the post-war status quo.[22] The European allies also reject the strategy's linking of their security to Soviet military actions in other areas, which implies a relegation of their national interests to the exigencies of superpower competition. Finally, it is most unlikely that West Germany would be willing or able to absorb and support the additional US forces which implementation of the strategy would require.

Thus a counter-offensive military strategy would be very unstable in a crisis, would require unattainable military force requirements and is solidly rejected by Allied political leaders as incompatible with NATO's charter.

DEFENSIVE MILITARY STRATEGIES

A number of European analysts have argued that the principal
source of insecurity in Central Europe is the fear of aggression on the
part of both Eastern and Western states, based largely on worst case
assessments of each other's military capabilities. They argue that the
strategies and force postures of both NATO and the Warsaw Pact
are offensive in character and highly unstable because they invite
pre-emption. These analysts believe the military preparations of
these states are in and of themselves part of the security problem.
This assessment has been taken up by several European political
parties. As the West German Social Democratic Party's (SDP)
August 1986 policy statement, *Peace and Security*, characterized the
problem:

> Security concepts continue to be dominated by fear and the threat
> and counter-threat of force . . . As long as armaments programmes
> and strategic planning are based on the worst-case assumption, no
> security problems will be solved, rather new ones will be created.
> The general feeling of being under threat, which is both a cause
> and a consequence of the arms race, can only be overcome
> through negotiated, inter-bloc security.[23]

These theorists see the situation in Central Europe as a particularly dangerous illustration of this problem. NATO and the Warsaw
Pact confront each other with what are seen as excessively large,
heavily armoured and highly mobile forces which are ideally suited
for offensive warfare. The military doctrines and training activities of
both sides' forces are characterized as similarly offensive in character, thereby exacerbating fears of hostile action. States on both sides
of the East–West divide have adopted strategies and force postures
which assume that the likelihood of military aggression by the other
bloc in a crisis is quite high. Thus, some analysts argue, a shift to
military capabilities and doctrines which left states structurally
incapable of conducting offensive military actions, or even the
replacement of traditional military forces with plans for civilian-
based resistance to any use of force, would stabilize European
security. Such a shift in defence postures would be reassuring to
neighbouring states but credible enough to guarantee a country's
security in the event that deterrence failed. Consequently this new
'defensive-only' posture is claimed to constitute a purer form of
deterrence by denial than now exists, since it would remove even the

possibility of punishment inherent in the structure of current NATO and Warsaw Pact forces. These concepts are the antithesis of the counter-offensive strategy.

Context

Concern about NATO's offensive military capabilities is not new and is widely shared in West Germany and among the left elsewhere in Europe. As Helmut Schmidt, hardly an advocate of these alternative defence concepts, wrote in 1962: 'The optimum goal of German defence policy and strategy would ... be the creation of an armaments structure clearly unsuited for the offensive role yet adequate beyond the shadow of a doubt to defend German territory.'[24] Contemporary advocates of a defensive, non-provocative defence believe that this goal has been obscured. In place of the current military structures, these theorists would field much smaller military forces structurally incapable of offensive operations, *strukturelle Nichtangriffsfähigkeit*.

Another rationale for eliminating all offensive military potential can be found in the SPD's proposal for an East–West partnership in the search for common security. As the SPD policy paper suggests: 'The alliance must make allowance for the political and military concerns of our eastern neighbours. It must, therefore, clearly articulate the defensive character of its strategy by placing the emphasis on border-area defence.'[25] For those who hold this view, the East and West face a common threat: the risk of war through miscalculation. The current politico–military situation is seen as an inadequate basis for stability and lasting peace. As such, the political institutions, force structures and military strategies of both alliances should be used to support the transition to a new security partnership, *Sicherheitspartnerschaft*. Arms control agreements that remove the threat of attack are seen as the most important instruments for developing this partnership. Ultimately, the new world political order should be based on agreed procedures for the peaceful settlement of disputes, and war should be proscribed as a means of achieving political objectives.

While there are many common themes in the alternative defence debate in Europe, there are two general schools of thought which have been characterized as the 'radical' and 'moderate' approaches. The former, which tends to be unilateralist in implementation, stresses the importance of reshaping the role of the West European states in NATO, of developing a non-nuclear Western Europe, and of changing the social structure. The more moderate version, reflected in the SPD's thinking, seeks to strengthen the European pillar of the Alliance, and envisages using bilateral arms control to achieve phased denuclearization of all of Europe and a shift in the ratio of standing armies to reserve forces in favour of the latter.[26]

Military Concepts

Most of these non-provocative proposals are based on concepts of territorial or area defence, in which the defenders seek to exploit the natural terrain and, in some instances, urban sprawl to wear down an aggressor. They also share an assumption that technology and history suggest defenders have decisive advantages over attackers.

West German interest in the desirability of non-provocative defence was stimulated in the 1980s by growing discontent with what are viewed as destabilizing trends in NATO and Warsaw Pact strategies. These critics of the military status quo roundly reject all notions of nuclear war-fighting and find the NATO operational sub-concept of non-nuclear attack of Warsaw Pact follow-on forces and the manoeuvre-oriented US Army doctrine, ALB, excessively offensive in character.[27]

Egbert Boeker and Lutz Unterseher have offered a clear definition of non-provocative defence: the build-up, training, logistics and doctrine of the armed fores are such that they are seen in their totality to be unsuitable for offence, but unambiguously sufficient for a credible conventional defence. Nuclear weapons fulfil at most a retaliatory role.[28]

Thus non-provocative defence implies a policy of no-first-use of nuclear weapons. Most advocates envisage the maintenance of limited numbers of nuclear weapons in Europe deliberately struc-tured to be unsuitable for first strike or counter-force use but capable of striking a variety of non-military targets and providing links to the American nuclear guarantee as a deterrent against first-use by the East.

Another common assumption of these concepts is that the most significant recent breakthroughs in military technology are those in surveillance, target acquisition and firepower, and that these ad-vances will give the defender a substantial advantage in any future conflict. Proponents argue that while mobility and armour have improved only marginally, there have been quantum leaps in the lethality of individual weapons and in reconnaissance capabilities.[29] These developments, it is contended, augment the natural advantages of the defender while reducing his vulnerabilities. That is to say, the defender can still fight from prepared positions on familiar territory, but can now allocate firepower remotely without having to manoeuvre forces into position. The defender no longer has to concentrate his own tanks and artillery to counter a massed attack because he can use precision-guided cluster munitions to achieve the same lethality.

In contrast, it is argued, the attacker's advantage of surprise has been reduced by improved surveillance systems. Incremental advances in mobility systems have not greatly enhanced his prospects of achieving rapid advances, for even helicopters fly at altitudes where they will be vulnerable to air defences. Finally, the attacker's need to defend himself has not diminished very much despite advances in armour.

Proponents of this posture advocate exploitation of some of the same technologies – new surveillance systems, PGM, RPV and air defences – that NATO commanders find attractive, but the non-provocative defenders propose to use them in a much more reactive fashion. Capabilities for deep forward strikes are eschewed because, even if procured for defensive purposes, their offensive potential still presents a threat to the Warsaw Pact. Such potential undermines crisis stability, which is perhaps the most important component in enhancing European security. NATO and Warsaw Pact conventional deep strike systems are seen as particularly destabilizing because their potential to diminish the other side's defence encourages pre-emptive attacks.

A few other common characteristics of these non-provocative defence concepts should be noted. Through the deployment of small, dispersed forces in less populated areas and abandoning the defence of urban centres, these proposals seek to increase crisis stability by providing few targets worth pre-empting before a war starts. For this reason, also, these concepts avoid or limit the role of air forces because airfields and other support facilities provide tempting targets for hostile aircraft and missile attack. Moreover, advocates emphasize that non-provocative defence would further increase stability by developing a new kind of military balance. This balance would not be measured in traditional 'bean count' fashion, but 'in terms of relative chances of successfully denying an aggressor his victory, without

calling destruction on the civilian population'.[30] Little attention has
been paid to naval forces by these theorists. The Danish Social
Democrats have advocated the replacement of the country's warships
with shore-based missiles for coastal defence, and with fishing boats
for surveillance missions.[31] Such thinking must assume that freedom
of navigation will be maintained by the US.

Four general types of new non-provocative defence concepts have
emerged in recent years: area defence (*Raumverteidigung*); wide
area covering defence (*raumdeckende Verteidigung*); the fire barrier
(*Grenznahe Feuersperre*); and integrated and interactive forward
defence. All of these concepts seek to deter aggression by denial. If
deterrence failed, defenders would attempt to ensnare any aggressor
in a web of small engagements, avoiding any decisive battle, thereby
precluding the achievement of a clear victory by the aggressor.

Horst Afheldt, one of the earliest proponents of this general
approach, has criticized NATO's current strategy of Forward
Defence because its requisite concentrations of forces are attractive
targets for pre-emption. Afheldt's thinking has been influenced by
Bogislav von Bonin, a West German defence planner who devised a
plan for non-provocative defence in depth of West Germany in the
early 1950s, and by the French military analyst, Guy Brossolet, who
argued that defence by attrition using light infantry (*armée de
couverture*) could be very successful even without a decisive battle
being waged.[32] Under his initial concept of area defence, Afheldt
would replace NATO's large heavily armoured units with static light
infantry, organized in 10 000 20–30 man formations. These small
units, armed with ATGW, would each defend 10–15 km^2 of territory
with which they would be very familiar.[33] In the forward area these
units would be comprised of active duty forces who would guard
against surprise attack. Rear areas would be covered by local reserve
units, or, in Afheldt's more recent 'transitional approach', by
traditional armoured units who would attempt to halt breakthroughs.

All these 'techno-commando' units would be both difficult for the
attacker to locate and unsuited for offensive operations. Any efforts
by an aggressor to concentrate forces against them would be
disrupted by precise fire from short-range artillery and rockets based
in dispersed patterns deep in rear areas. Afheldt has argued that
these light infantry units, armed with their ATGW and knowledge of
the terrain and supported by short-range rocket and artillery
batteries, all linked together by an integrated communication

system, could be effective in depleting the size and momentum of a Warsaw Pact offensive close to the inner German border. Thus Afheldt's concept relies heavily on emerging conventional weapons technology. It is also premised on the maintenance of an American nuclear guarantee in the form of SLBM.

A number of similar concepts have gained some political support in Europe. Former *Bundeswehr* Major-General Jochen Löser has proposed a 'wide-area territorial defence'.[34] This concept envisages the establishment of a frontier defence zone 80–100 km deep in which barriers and blocking units channel attacking tank forces towards concentrations of fire. Löser advocates the deployment of a network of light infantry 'shield' brigades in the forward zones to wear down an attacker in a series of small engagements. These 'shield' units would cooperate with traditional allied and German armoured and mechanized units in the transition period. However, once the necessary force restructuring process was completed, light units would comprise the bulk of both the 'shield' and the 'sword' forces. Löser's scheme would require a doubling of the number of brigades in the *Bundeswehr* by expanding the reserves. He would maintain the bulk of these units in the second echelon for counter-attack against breakthroughs. Löser's concept would make extensive use of air and missile defences, and does not rule out the ultimate use of nuclear weapons.

Norbert Hannig and Albrecht von Müller have advanced some similar ideas for non-provocative, high-technology, forms of forward defence. Von Müller has refined a number of the concepts under discussion into a plan for a four-layered 'integrated forward defence' (IFD).[35] The first layer would be a 'fire-belt' extending 5 km to the west of the inner German border but 40–60 km to the east in the event of hostilities. No NATO troops would be deployed in this *cordon sanitaire*, which would be capable of being inundated with remotely delivered fire from MLRS rockets, artillery, mortars and intelligent mines. Successive defence layers would include: a 'net zone' of 20–50 km, covered by small units of light infantry which would conduct hit-and-run attacks with PGM against advancing forces; a 'manoeuvre zone', where heavily armoured but dispersed units would conduct operations to halt any forces that penetrate the second zone; and a 'rear defence zone', essentially the balance of NATO territory, which would be defended against air-mobile and special operations forces by a network of local, semi-mobile, territorial defence units.

Von Müller's concept is one of the few to address the air war. IFD would include some counter-air operations and close interdiction against choke-points. However, von Müller notes that these missions will be best performed in the future by rockets and cruise missiles, thereby freeing NATO air forces for air superiority missions over Allied territory. Thus his plan addresses the criticism, advanced with respect to many territorial defene schemes, that it is not in West Germany's interest to absorb the brunt of a Warsaw Pact offensive deep inside its borders and then attempt to eliminate it.

Hannig similarly proposes the development of an uninhabited, 4 km-wide 'fire barrier' along the inner German border, which could be saturated with fire from mobile surface-to-surface and surface-to-air rocket and missile launchers deployed at various ranges behind this zone. This concept envisages inflicting heavy losses upon any aggressor as soon as the border is crossed, thereby breaking the momentum of an attack.[36] Anti-armour units equipped with PGM would operate just behind the fire barrier to support the attrition process and deal with any breakthroughs. To cope with defence of the rear against airborne landing forces, Hannig would simply redirect fire through 180 degrees from the same rockets that constitute the fire barrier. This aspect of Hannig's scheme has prompted the alternative defence community to dub his proposal FOFA in reverse. Hannig essentially proposes to accomplish the counter-offensive and interdiction missions presently performed by armoured manoeuvre forces and fighter-bombers with conventionally armed missiles of varying ranges.

The West German Study Group on Alternative Security Policy has advanced a non-provocative defence proposal called 'interactive forward defence' which integrates elements of several of the concepts described above. The proposal has three components: a static 'containment force' composed of decentralized light infantry units employing reactive tactics; a 'rapid commitment force' comprised of mechanized infantry, armour and cavalry forces with limited mobility; and a 'rear protection force' to cope with penetrations and airborne assaults. The first two components would be comprised of active duty NATO units, with the latter filled out by reserve forces. As the chief theorist of the Study Group, Lutz Unterseher, explains, the static warfare units would maintain area control and deplete an adversary's momentum by harassing advancing units and channelling them into areas where they would be vulnerable to attack by the mechanized forces.[37]

The non-offensive character of this force would be evident from the size of the units. The mechanized forces would total only 180 000 troops, with the bulk of NATO's wartime strength consisting of 500 000 West German troops organized into these new static infantry units. Moreover, the mechanized troops would be supported by a decentralized, static system of logistics protected by the containment force, which would restrict their movements to ranges within the static system. The Study Group favours these mobile forces in the forward zones of the static defence as much less costly substitutes for the area saturation by rockets and missiles that is an essential element of the other proposals. They contend that adoption of their tactics would allow for a *Bundeswehr* with one-third fewer troops and two-thirds fewer armoured vehicles to halt a Warsaw Pact attack close to the border.

The military thinking and political goals of these German strategists are echoed in a number of other European proposals such as those of the British 'Just Defence' organization and The Alternative Defence Commission.[38] In advocating a denuclearized posture of 'defensive deterrence' decoupled from US strategic forces, the latter group concedes that the political imperatives of Forward Defence require NATO to retain or even expand its large standing armies, but argues that more systematic preparations for territorial defence should also be explored as an alternative.[39] All these alternative models seek to reduce Eastern perceptions of a military threat by fielding military forces that offer few targets for pre-emption and are structurally unable to attack.

Non-provocative defence proposals are attracting growing political support from the European left. The West German SPD's 1986 defence paper endorsed a restructuring of the *Bundeswehr* that would 'render it incapable of mounting offensive operations' but increase its capability to support Forward Defence. This restructuring would emphasize the army's border defence, containment and air-defence functions with the stated goals of enhancing stability, facilitating crisis management, and reassuring West Germany's 'eastern neighbours' of NATO's defensive intentions.[40] Other European socialist parties have endorsed the general notion of a more defensive posture for NATO. An October 1986 British Labour Party policy document called for restructuring all branches of the UK's armed services towards defensive military roles, including conversion of the Royal Navy's attack submarines to coastal defence missions.[41]

The aforementioned proposals for non-provocative defence have very doubtful prospects of providing effective deterrence because they falsely assume that the problem faced by the attacker and defender are totally different. Most of these concepts leave the defender with at most only very limited capabilities for the offensive counter-attack which may be necessary to repulse or evict invading forces. A more effective deterrent would blend some elements of punishment and denial. By relying solely on the pure form of deterrence by denial inherent in non-provocative defence concepts, a state runs the risk of tempting a potential aggressor to wear down its defences by a sustained attack. A related operational shortcoming is that these concepts advocate largely reactive measures to be undertaken only after an invasion, making them highly vulnerable to surprise attack. They also rely very heavily – arguably excessively – on reservists, whose standard of training would at best be variable, for initial defence.

Most of the current European concepts, particularly Afheldt's, are excessively focused on the threat of heavily armoured assaults. The NATO countries and most industrialized states confront multi-dimensional security threats, including combined air and ground force operations. According to a study by the Dutch Ministry of Defence, the sorts of anti-tank oriented defence postures proposed in these concepts could be severely degraded by infantry attacks supported with artillery fire.[42] The ATGW units may also be subject to piecemeal destruction by heavy mechanized forces. Such defences could also stimulate and be overtaken by a technical arms race in ATGW counter-measures and new forms of armour, particularly the reactive armour already reported to be available for Soviet T-64B and T-80 tanks.

Finally, all of these concepts are premised on the realization of some immature conventional weapons technologies and on the debatable assumption that these emerging technologies are shifting combat advantage to the defence.

Nevertheless, systems analysis studies in Germany have suggested that at least some of these other possible defence options are more cost-effective than some of the *Bundeswehr's* existing Active Defence plans. Several analysts involved in these studies have argued that traditional and alternative defence concepts should be considered in a more integrative, rather than a mutually exclusive, fashion. Indeed, one such study concluded that the 'incorporation of properly designed reactive defence into NATO's existing force structure could indeed contribute to a significant improvement of

NATO's forward defence at acceptable costs', even without relying on emerging technologies.[43]

The advocates of defensive defence have illustrated how their ideas would operate in the countryside, but have shown surprising neglect of how these concepts would be applied in urban settings. Instead, most have simply stated that they would avoid, as much as possible, conducting defensive operations in cities. This will be a particularly difficult feat in a country with West Germany's development density. This lacuna may be partly explained by political sensibilities in West Germany, where discussion of urban warfare is hardly popular. Those theorists who have broached the urban problem, such as Wilhelm Nolte, have suggested that civilian resistance in the cities would be the most appropriate adjunct to territorial defence concepts.[44] However, given the nature of the landscape in Central Europe and most modern industrialized states, urban warfare is likely to be a significant aspect of future conflicts. This situation calls into question one of the principal dividends touted by advocates of non-provocative defence, namely avoidance of the massive societal destruction which is likely to accompany modern conventional warfare.

All these theorists agree that the probability of war arising from miscalculation would be greatly reduced even if the West adopted a defensive defence posture unilaterally. Moreover, they argue that if the resulting deterrent did fail it would be much easier to achieve war termination with less capable forces on one or both sides than presently exist. In a similar vein, it is argued that the East's incentives to pre-empt NATO militarily would virtually disappear if the West had no capability to assume the offensive. It is contended that NATO's current offensive nuclear doctrine and any offensive conventional military plans or capabilities provide a convenient rationale for the sustained Soviet build-up. Finally, most advocates of this posture feel that it could, in the long-run, facilitate the evolution of a new politico-military situation in Europe, and may be essential to ending the division of the continent.

The moderate proponents of non-provocative defences advocate multilateral negotiations to achieve the desired force structures. The more radical advocates would pursue these changes unilaterally, while offering unspecified 'incentives' to the East to emulate this shift in its own military posture. It is hard to envisage what incentives the East would have to reduce its military edge in Europe after the West had made such a dramatic shift in its posture unilaterally.

Similarly it is difficult to imagine the circumstances, other than decline, under which the US and the USSR, with their global interests, would adopt a defence posture incapable of projecting power. Thus it is very likely that European nations would have to adopt any such concepts on a unilateral basis. Such a move might be feasible, although highly risky, in a world where nuclear weapons are still in place to provide an ultimate deterrent; the more the nuclear element in the theatre military equation is reduced the greater the hazard implicit in any move towards less comprehensively capable defences.

8 Twin Challenges

Whatever strategic concepts or conventional defence options they may choose to pursue through force improvement plans, NATO leaders will have to contend on the one hand with domestic demands for concomitant progress in arms control and, on the other, with increasingly sophisticated Soviet military and political counter-measures. The Warsaw Pact is acutely concerned with the implications of emerging developments in NATO doctrine and weaponry, and has already established the broad outlines of an integrated political and military response. Even before NATO's renewed attention to its conventional defences, the Warsaw Pact was engaged in a series of initiatives designed to improve the capabilities of its forces to manoeuvre and strike deep into NATO's rear areas. In addition, the Soviet leadership under Gorbachev has shown remarkable skill in its public diplomacy towards the West. In the wake of the INF treaty, Moscow seems sure to advance further dramatic proposals for sharply reducing battlefield nuclear weapons and conventional forces, such as those put forward in the June 1986 Warsaw Pact 'Budapest Appeal' and elaborated in the May 1987 Berlin Statement. The nature and credibility of the Alliance's response to Moscow's overtures will be critical. NATO governments cannot realistically expect to obtain domestic support for conventional force improvement plans without parallel efforts towards arms control. Thus, dealing with these Soviet military and political activities and shaping effective, broadly-supported military and arms control initiatives will be the twin challenges to NATO as it pursues its conventional defence options. This chapter offers a broad overview of these two complicated sets of issues.

SOVIET REACTIONS AND CAPABILITIES

The Kremlin is alarmed that NATO's new high-technology weapons, coupled with changes in operational concepts, could disrupt Soviet military plans developed over the past decade. Soviet analysts are debating how best to respond. Both military and political leaders have made statements which suggest that they are deeply concerned about the implications of NATO's new weapons technology and tactics. It is clear from Moscow's public reactions to General Rogers' statements about his conventional defence goals that the Soviet Union will not allow NATO to realize these new military capabilities without a sustained diplomatic offensive bolstered by countervailing military responses. Indeed, in a lengthy piece in *Krasnaya Zvezda* reviewing the principal NATO weapons systems required for second echelon targeting and the enhancement of firepower, a prominent general noted, 'The Soviet Union cannot remain on the sidelines in the presence of this danger'.[1] In the short term, Moscow hopes to encourage opposition to conventional defence modernization programmes in the West while both deploying counter-measures to new NATO deep-strike systems and developing similar weapons of their own.

Moscow has engaged in a sustained conventional and tactical nuclear build-up over the last decade designed to ensure a quick conventional victory and to deny NATO its nuclear escalation option. Whether or not the Soviets can maintain their steady growth growth in defence capabilities at a time when other spending priorities are receiving new attention is unclear. On the demographic side, the Soviets face a slightly different problem from the West. Because of relatively low birth rates in the European USSR, the Soviets will increasingly have to fill the ranks of the Army with youths from Central Asian republics, which are experiencing substantial population growth.[2] This demographic trend poses operational and political problems for the Red Army. Even if ethnic tensions do not begin to strain military cohesion, there are certain to be new logistical problems. For example, educational levels in the Asian republics are lower than in the Western USSR and at least two-thirds of the non-Russian population is not fluent in Russian. These are not ideal conditions for development of a high-technology army with less rigid command and control.

Over the past decade the Soviet military has developed plans and capabilities for conducting a protracted conventional war in Europe.[3] It is clear that the Soviet military leadership believes that nuclear weapons now have diminished military utility but that qualitative improvements in conventional means require new operational plans.[4] There is a broad consensus among Western analysts that the primary purpose of Soviet INF had become deterrence of NATO's use of corresponding nuclear systems. The non-nuclear offensive strike capabilities of Soviet ground and air forces have increased significantly during this period, as have their ability to locate and destroy

NATO tactical nuclear weapons and other high-value targets by conventional means.[5]

While the INF Treaty requires dismantling of the SS-23 and SS-12 mod the Soviet Union can continue to deploy in Eastern Europe new short-range missile systems (those with ranges up to 300 miles), such as the SS21, and new aircraft, such as the SU24 *Fencer* and MiG 31 *Foxhound*, all of which can be used to strike NATO rear areas with greater effectiveness than earlier systems. The Soviets have reportedly developed conventional submunition packages for the SS-21.[6] If these missiles are, or can shortly become, as accurate as current Western estimates suggest, they will provide Moscow with effective conventional deep strike options against NATO deterrent forces and reinforcement potential by threatening C^3I assets, air defence sites, airfields and key choke-points.

In response to the lessons of the 1973 Middle East War and NATO's deployment of improved anti-tank weapons, the Soviets have significantly improved the inherent protective capability of their tank armour and increased both the flexibility of their armoured forces and their ability to conduct rapid, large-scale attacks. The Soviets have restructured their fronts to provide improved air, artillery and mobile infantry support for armoured attacks. They have revived a combat organization known as the OMG. The OMG is a task-oriented force that allows for more effective integration of combined arms. OMG are designed to penetrate NATO defences in quick thrusts. The enhanced mobility and firepower of Soviet general purpose forces pose an increasing threat to NATO's ability to defend forward.

Development of NATO capabilities for conventional deep strikes against mobile forces, improved air and missile defences, and enhanced warning and surveillance would undermine Warsaw Pact gains from their sustained force improvement programme. Indeed, some Soviet military writers have suggested that increasingly lethal NATO conventional weapons would slow Soviet rates of advance in the initial stages of war.[7] Thus the success of Soviet *blitzkrieg* operations could be in doubt. Former Chief of the Soviet General Staff, Marshal Nicolai Ogarkov, noted another concern:

The sharply increased range of conventional weapons makes it possible immediately to extend active combat operations not just to the border regions, but to the whole country's territory . . . This qualitative leap in the development of conventional means of destruction will inevitably entail a change in the nature of the preparation and conduct of operations.[8]

In other words, the Soviet Union would also have to take into account defence of their rear areas and C^3 assets against more effective conventional attacks.

Soviet military writers have suggested two general types of response to NATO's conventional challenge. One group argues that *blitzkrieg* operations can still work if they are employed with still greater speed and stealth. Another school sees *blitzkrieg* operations increasingly threatened by NATO's emerging conventional weapons. The latter argue that the Soviet Union and its allies must now prepare for a protracted conventional or possibly conventional/nuclear war in Europe and that the Soviet Army must consider manoeuvring in more dispersed formations.[9] Under such circumstances, surprise can be achieved by rapid concentration to penetrate and exploit weaknesses in enemy lines, in particular by the bold use of task-oriented OMG. This tactic was actually tested during the Warsaw Pact *Zapad '81* exercise. There is some evidence that the Soviet Union is already restructuring its forces for these kinds of operation.[10]

If Soviet forces are moving towards this posture and away from rigid echelonment along breakthrough corridors, then SACEUR's FOFA concept of disrupting the movement and timing of second echelon forces may have a more limited effect on the course of battle than envisaged by NATO planners. Under this new Soviet concept, forces from the second echelon might be more predominantly involved in occupation of territory and less critical to the achievement of a breakthrough. Another way for the Soviet Union to reduce the threat to its reinforcement capabilities posed by FOFA might be to move additional divisions into garrisons in East Germany and Czechoslovakia. Although it would be difficult to reconcile such a move with their proposals for conventional arms control and the reduction of tension, Moscow could arguably characterize this action as a purely defensive precaution against 'NATO's new offensive strategy'. Thus there is no reason to assume, as some Western strategists appear to, that Soviet doctrine will not find ways of countering various innovations by NATO.

With regard to NATO's use of new surveillance and target-acquisition systems, it should be recalled that the Soviet forces have a highly-developed doctrine of radio-electronic combat (REC) and deception (*Maskirovka*). Fairly simple counter-measures, such as radar corner reflectors and absorptive materials, are already available and more are certain to be developed which can confuse the target acquisition and precision-guidance systems critical to the success of deep strikes with conventional munitions, and even those for shorter-range PGM. In addition, there is concern that the new reactive armour deployed on Soviet T-64B and T-80 tanks will defeat many of the Western munitions, including some PGM and ATGW.

NATO capabilities for deep strikes are liable to become high priority targets for Warsaw Pact forces and could even provoke pre-emption. Indeed the Soviet Union might decide to use nuclear weapons in order to have high confidence of destroying these mobile NATO deep strike weapons and other critical targets, such as command posts, deployed well behind the front.[11] The Achilles' heel of all these new NATO concepts is the C^3I facilities where targeting data is assembled, processed and transmitted to the battlefield commander. Disruption of even a few critical nodes in this system could greatly reduce its effectiveness.

Moscow's Political Offensive

The central elements of Moscow's political response to deep strike concepts and other conventional force improvement initiatives are already well established. Playing to the fears of many West Europeans, the Kremlin has characterized INF missile deployments and the FOFA concept as part of an American desire to confine any war that breaks out in Europe to that continent and to avoid retaliatory strikes on the US. Soviet observers also argue that the *sine qua non* of the Rogers Plan is the attainment of a fundamental shift in the correlation of general purpose forces which would give NATO military superiority and the ability to defeat Warsaw Pact forces on their own territory. They allege that General Rogers and others who advocate enhancing conventional forces as a means of reducing NATO's reliance on nuclear forces are engaged in a great deception, and that the planned improvements are actually designed to supplement rather than replace Allied nuclear potential.

Reflecting the classic 'carrot and stick' approach that was used throughout the INF debate, the Soviet Union has advanced seemingly dramatic arms limitation proposals in tandem with threats of countermeasures. Since the 1983 Prague Political Declaration, the Warsaw Pact has declared its willingness to negotiate limitations on all types of weapons systems in Europe on the basis of 'equality and identical security'. This approach was refined in the Sofia Statement of 24 October, 1985. In this document, members of the Warsaw Treaty Organization called on the US and the USSR to agree not to develop or produce 'new types of conventional arms that are comparable to weapons of mass destruction in terms of their destructive possibilities'.[12] Further insights into the East's approach have emerged from the June 1986 Budapest Appeal, the May 1987 Berlin Statement, and pronouncements by General Secretary Gorbachev and Prime Minister Jaruzelski of Poland.[13] The Warsaw Pact nations have declared their interest in an integrated, phased approach to future European arms control negotiations beginning with mutual 25 per cent reductions in conventional forces and armaments, in tandem with elimination of short-range nuclear forces, by the early 1990s and followed by even deeper force cuts by the year 2000. During the first stage, the Pact proposes to reduce the risk of sudden attack through 'reciprocal withdrawal of the most dangerous, offensive types of weapons from the zone of direct contact of the two military alliances'. The Eastern states have endorsed the notion of eliminating, 'in the course of reduction', the disparities in NATO and WTO forces that have evolved over time. Eastern officials also favour a tradeoff between reductions in Eastern tanks and reductions in NATO aircraft and helicopters.

Various Pact officials have stated that the time has come to rectify what they now admit are asymmetries in Warsaw Pact and NATO forces through arms control. It must be noted, however, that because the official Eastern assessment of the balance is so different from the range of nearly all Western estimates, it is hard to know exactly what levelling actions this would require. A 1987 Soviet publication contended that the 'combat capability of the NATO armed forces roughly matches that of the Warsaw Treaty'.[14] The Soviet Union argues that NATO and the Warsaw Pact have equal numbers of troops. While the Warsaw Pact countries admit to having more tanks, they argue that NATO has a countervailing advantage in anti-tank weapons and more 'battle-ready' divisions, attack aircraft, and helicopters than the Pact. Thus it is very possible that future negotiations will be hindered, as was MBFR, by East–West disputes over data and units of account.

Ominous pronouncements are certain to emanate from the Kremlin when any NATO conventional force improvements get under way. Vague warnings about the dangers of new 'offensive' military capabilities, which have been appearing 'once again' in West Germany, might be coupled with revelations of countervailing Warsaw Pact military actions. Such a propaganda offensive could be effective in exacerbating what could easily become a heated debate in several NATO countries.

Thus the Soviet Union and its allies are not without a number of emerging conventional defence options of their own which may both present new threats to the West and counter the effectiveness of a number of NATO initiatives. In addition, Moscow has made it clear that it will wage a vigorous political and diplomatic struggle to complicate NATO's achievement of significant new conventional military capabilities. These complications, coupled with the economic and demographic constraints discussed earlier, make it clear that the Alliance has a difficult road ahead in improving the conventional component of its deterrent posture.

THE ROLE OF ARMS CONTROL

This leads to a consideration of the potential contribution of arms control to improving the conventional balance and of the complications which some military initiatives may create for arms control, both nuclear and conventional. The development of a parallel arms control strategy has become an essential element of any NATO force modernization plan. While the dividends are likely to be modest militarily, sustained and genuine arms control efforts will be politically indispensable.

Members of the North Atlantic Alliance are firmly committed to enhancing security in Europe through the pursuit of bold new steps that would result in greater openness in military activity as well as a comprehensive, stable, and verifiable balance of military forces at lower levels. This position, first enunciated in the Halifax Statement, was reaffirmed in the Brussels Declaration and subsequent NATO documents.[15]

It is unlikely that conventional arms limitations would greatly reduce Western military requirements, given the demands of NATO's forward defence strategy and the relative ease with which any Soviet forces or equipment withdrawn could be reintroduced into a Central or pan-European zone of reductions. In contrast, any US forces withdrawn from Europe would have to return from bases across 3000 miles of ocean. Preliminary analyses suggest that very large and highly asymmetrical reductions of Warsaw Pact forces *vis-à-vis* NATO armies in the overall Atlantic to Urals zone, with particular emphasis on the old MBFR Guidelines Area, could enhance the effectiveness of NATO's projected conventional military capabilities in blunting a Warsaw Pact offensive, assuming the availability of timely and unambiguous warning of mobilization and political preparedness to react to it.[16] In this regard, the stabilizing effect of any arms control arrangements in Europe would be enhanced by transparency measures that improve NATO's understanding of Eastern military operations and provide warning of Pact mobilization.

Under the Long Term Defense Programme NATO pursued improvements to both its conventional and nuclear forces. The tenth element of this initiative called for the modernization of NATO's nuclear posture. In addition to the deployment of the 572 LRINF (longer-range intermediate range nuclear force) weapons, NATO ministers agreed to reduce Allied deployed warhead totals by 1000. NATO military planners advocated the new deployments as a way to achieve a more effective threat of retaliation on the Soviet homeland and a more balanced mix of forces, then heavily weighted towards highly-vulnerable short-range systems. At the Montebello ministerial session in December 1983, NATO took a further unilateral step towards the development of a balanced nuclear posture by deciding to withdraw another 1400 short-range weapons. This programme resulted in an overall reduction of NATO's nuclear arsenal from some 7000 weapons in 1979 to some 4600 today.[17]

In pursuing this restructuring in the wake of the INF Treaty, the notion of attempting to achieve a partial trade-off between NATO's battlefield nuclear weapons (BNW) and the Warsaw Pact's 2.5:1 advantage in troop strength in the Central Region has become politically attractive in some quarters. NATO offered such an arrangement to the Warsaw Pact during the 1970s, at a time when the Pact had many fewer BNW deployed in Europe, as part of the so-called Option III package in the MBFR negotiations.[18] Advancing this proposal, some argue, would create an explicit link between NATO's deployment of BNW and the presence of so many Soviet heavy armoured divisions in Central Europe. Such a plan should be avoided until there is significant progress on conventional arms control. Even substantial Warsaw Pact troop withdrawals along the Central Front would not necessarily eliminate the potential for these forces to be redeployed in a period of heightened tension, nor would it eliminate the other mission of NATO's BNW force, the deterrence of military aggression, particularly through demonstrating a plausible threat of the use of nuclear forces in implementation of the flexible response strategy. The establishment of parity in critical elements of military capabilities would be one way to induce greater stability. Equal limits on equipment that is particularly well suited for offensive operations, such as tanks, long-range artillery, and obstacle-crossing equipment should be explored. There is little doubt that a common ceiling on tanks and artillery, or other equipment such as armoured fighting vehicles, well below the current levels deployed by the Warsaw Pact could remove a great deal of the punch from both a short-warning and a mobilized attack. However, limitations on one or two categories of weapons systems do not necessarily remove the Warsaw Pact's offensive threat that comes from the interaction of a number of force elements in their combined arms strategy. For example, if an accord that mandated parity in tanks were followed by an unbridled competition in attack helicopters and artillery, imbalances in these weapons might well become the principal sources of military instability in Central Europe. It would be possible for an accord to simply change the configuration of offensive military power. Thus it is clear that structural limitations, such as force ceilings, should be carefully integrated with one another and with operational constraints if arms control is to yield enduring stability. Moreover, any limitations on manpower or equipment in Europe will require agreement to a much more intrusive verification regime than has ever been achieved before.

Operational arms control measures designed to make military activities more transparent, thereby reducing concerns about surprise attack, might enhance stability in Europe more than some force reductions. Inspections and even more extensive activity notification requirements than were achieved in the Stockholm CDE could make clandestine mobilization and/or other preparations for attack very difficult to achieve. The Stockholm Agreement of 1986 established a useful foundation for further progress in redressing some of these central threats to peace in Europe. Strict observance of the Stockholm Agreement will alleviate, but by no means eliminate, uncertainty about the timing, scope, and nature of military activities. However, much bolder steps, including both reductions of asymmetries in force levels and measures that would give greater predictability and transparency to military activities, will have to be implemented to reduce significantly the risks of war in Europe.

Overall, NATO must cope with a dynamic Soviet political and military challenge and simultaneously develop a meaningful set of arms limitation proposals that can be pursued as a complement to any conventional defence improvement initiatives. It must also face the fact that any significant progress on conventional arms control, or even the promise thereof, is likely to be accompanied by some erosion in domestic support for strengthening NATO's general purpose forces. As recent history has shown, maintaining support for defence programmes in a period of improving East–West relations requires careful public education about the limits of *détente* and the continuing need for prudent military planning.

9 Conclusion: Improving NATO's Conventional Defences

Increasing the non-nuclear component of Western deterrence, even marginally, will be a long and complicated process. There are no technological or political shortcuts. Financial and demographic constraints will require that careful trade-offs be made in resource allocation. In the short term, NATO should attempt to realize incremental improvements to its capabilities for initial defence, counter-air operations and the disruption of Warsaw Pact reinforcement potential. NATO defence planners generally favour these priorities and the European Allies will support a programme of this kind provided they are convinced that the specific measures pursued will enhance crisis stability and not diminish linkage to the American nuclear guarantee. The expanding cooperation in weapons procurement and defence planning by the European members of the Alliance has the potential to yield significant military dividends and economies in resource commitments. The various ongoing and planned arms control negotiations offer some modest opportunities for enhancing crisis stability and improving the general political climate.

This chapter proceeds from the assumption that the conventional balance in the Central Region is not so adverse that modest gains cannot make a difference in NATO's ability to cope with many aspects of a Soviet attack and therefore strengthen deterrence. Despite the Warsaw Pact's quantitative lead in force levels and deployed weapons systems and growing technological parity, NATO's conventional forces in the Central Region already provide a strong deterrent to Soviet aggression and would put up a very formidable initial defence. Nevertheless, NATO's current military posture does not instil high confidence that the Alliance could avoid relatively early resort to first-use of nuclear weapons to halt an offensive by larger, more rapidly reinforced Warsaw Pact ground forces. Extrapolation from current trends clearly indicates that the balance is shifting in favour of the Warsaw Pact.[1] However, NATO can slow and reverse this shift within prevailing economic and political constraints by the selective application of new technology,

131

operational innovations, and greater defence cooperation.

Given the fragile political and economic base upon which any programme to shore up the conventional component of Western deterrence will have to be constructed, careful long-range planning and consensus-building within the Alliance is required. NATO needs to strengthen its long-term planning mechanisms to harmonize and give direction to various national efforts and to maximize the military output of defence expenditures. It may well prove desirable to retain and enhance certain features of the LTDP, particularly the independent monitoring of programme implementation, in the CMF process. In addition, the CMF process should be used to continue to clarify and update NATO's strategic concept for conventional defence.

The US can assist and enhance Allied defence collaboration by continuing to develop the two-way street in armaments trade by such programmes as that initiated by the Nunn Amendment. At the same time European governments must realize, as shown by the case of the UK *Nimrod* AEW programme, that buying American technology will often be more cost-effective than going it alone.

Expanded Franco–German defence cooperation is critical to strengthening NATO's conventional capabilities, and that is why this book has devoted considerable space to the issue. However, there is little doubt that the combat capability of the French Army will diminish. French plans call for programmed reductions in manpower to be compensated for by an increase in mobility and firepower, but budgetary constraints are likely to limit the replacement of outdated equipment. It is unlikely that Paris will abandon its strategy of preserving maximum flexibility of action and declare precisely when and how its conventional and nuclear forces would be used in defence of West Germany. Nevertheless, France's apparent new interest in greater coordination with NATO plans should be exploited to the full, particularly with regard to taking specific measures that would prepare its logistic base for support of Forward Defence. The expanding scope of Franco–German defence planning may provide increasingly significant military dividends and, in the process, make West Germans feel somewhat more secure. In this connexion the creation of a joint Franco–German brigade, while of questionable direct military value, may be of substantial symbolic importance.

Other elements of European defence cooperation can only contribute to the vitality of NATO. The discussions in the WEU and other bilateral and multilateral forums designed to coordinate policy development and military plans are already yielding dividends. These initiatives are likely to bolster European commitment to conventional defence improvements and to enhance European influence in Allied policy deliberations. So far as specific conventional defence improvement measures are concerned, there are several practicable and relatively affordable steps which NATO could take to improve its ability to deal with the leading elements of any Warsaw Pact offensive. It has started with relatively low-technology measures such as increasing munitions stockpiles and stocks of prepositioned war reserve equipment, sheltering aircraft, improving capabilities for receiving and deploying reinforcements, and enhancing the readiness and training of reserves. Further steps need to be taken to rectify the peacetime malpositioning of in-place forces and to expand NATO's air defence and counter-battery capabilities. In addition, NATO should carefully examine how it can make more effective use of barriers. There have been significant advances in non-obtrusive barrier technology which could be employed in ways that would circumvent long-standing political objections to peacetime barrier preparations. For example, buried pipes have been developed which can be filled in a crisis with liquid explosives and detonated to create tank traps.[2] NATO could also look at ways of making better use of the barriers created by the natural and urban landscape.

NATO needs to develop a survivable, theatre-wide, all-weather C^3I system, which can incorporate information from US (and possibly French) satellite systems and which is resistant to Warsaw Pact electronic warfare measures. While maintaining the integrity of its own C^3I system, NATO should develop more effective capabilities to disrupt Warsaw Pact *tactical* C^3 as an additional means to destroy the cohesion of its rigidly-controlled offensive operations. It is, however, doubtful whether (even if practicable) it would be advisable for NATO to attempt to develop the capability totally to neutralize Warsaw Pact theatre C^3I because this might be so threatening as to force rapid escalation and reduce the chances of attaining a negotiated settlement.

Overall, emerging technology looks less promising than it once did. Its impact on the nature of warfare will probably be evolutionary rather than revolutionary. None the less there are a number of affordable emerging and emerged technologies which could enhance NATO's front-line defences in the near term. These include MLRS, a number of precision-guided weapons, and several surveillance and target acquisition systems. MLRS is a good example of an ET system which can be brought into operation quickly at key points of the battlefield and utilized effectively without the need for an extensive C^3I support system. It has the added dividend of already being procured on an Alliance-wide basis, and it can be progressively upgraded with developments in target acquisition capabilities and submunitions. In the coming years, NATO should concentrate its resources on programmes like this which have short-term pay-offs for improving forward defences.

There is no requirement for any sweeping change of NATO doctrine in order to enhance crisis stability or to reduce the risks of rapid escalation on the Central Front in Europe; and neither would such a change be desirable. The threat of nuclear retaliation, whether under the earlier doctrine of Massive Retaliation or the current doctrine of Flexible Response, appears to have established and maintained its deterrent value over the past three decades. While it is impossible to prove that deterrence has been the prime factor, there has at least been no war in Europe over the period. No purely conventional deterrent could raise the risks and costs of going to war so dramatically. Most Europeans show little interest in relying on a largely or exclusively conventional deterrent, given the fact that such postures have so often failed.[3] The existence of nuclear weapons inhibits both superpowers from contemplating the use of force to achieve political objectives in Europe. Assertions about the declining credibility of nuclear deterrence in an age of superpower parity are overstated. While the probability that NATO would use nuclear weapons is generally judged to have declined, even the possibility that they might be used imposes great caution on the Soviet Union. Moreover, the extensive Western plans to use nuclear weapons and the pivotal geostrategic importance of Western Europe to the US would seem to reinforce this caution among Soviet leaders.

Nuclear weapons provide certain military as well as deterrent advantages. For example, under operational circumstances, the threat of nuclear strikes can be expected to keep the two-thirds to three-quarters of Warsaw Pact ground forces that are in the second echelon dispersed and less well positioned to exploit breakthroughs. It is most unlikely that improved conventional munitions will be able to substitute for the effects of nuclear weapons in this respect or in the performance of many other missions. Nevertheless, NATO should continue to thin out its battlefield nuclear inventory to the greatest extent possible consistent with its strategy, and take measures to improve the survivability of remaining missile and air-delivered weapons. Following on the 1979 'Dual Track' and 1983 Montebello decisions, NATO's nuclear posture was already evolving towards a more prudent mix of systems.[4] In the aftermath of the INF accord, programmes to safeguard short range missiles and dual-capable aircraft and artillery systems from Soviet pre-emption and to preserve a sufficient nuclear capability to maintain a viable Flexible Response strategy will be critical.

While a shift in doctrine may not be necessary, the strategic concept guiding NATO defence planning and training should be clarified. As has been argued in Chapter 7, maintaining capabilities to fight a more protracted conventional war (without specifying precisely how long this period would be) is the concept that would best enhance deterrence, enjoy the broadest political support, and probably be achievable within projected resource constraints. Planning and preparing for a strategy of no-early-use does not reduce the deterrent value of nuclear weapons, and neither does the possibility of ultimate nuclear use undermine Allied resolve to wage a vigorous conventional defence. This concept provides NATO leaders with a wide range of military options appropriate to the level of aggression and offers the best prospect of securing sufficient time to seek a resolution of any conflict at the lowest level of damage; ideally before any nuclear weapons are used.

It has been shown in Chapter 7 that the European members of NATO have no interest in planning for an indefinite conventional defence, even if it were economically feasible. Most Europeans fear that such a posture would weaken deterrence and have unacceptable consequences – the devastation of the continent – should deterrence fail. It is for this reason as well that concepts envisaging a denuclearized non-provocative defence are unlikely to attain much wider support in Europe than they currently have.

The notion that small units of light infantry armed with ATGW and supported by artillery barrages could arrest a heavily armoured Soviet offensive and then wear down penetrating forces is not supported by military history or current operational analysis. It takes large, tank-equipped units to stop such offensives. The non-provocative defence concepts also have at best only a dubious deterrent effect because of the limited range of counter-attack options available to the defender. Moreover, NATO confronts a multidimensional threat from the Warsaw Pact, including combined air and ground operations, and it is very likely that defences oriented to anti-tank missions would be highly vulnerable to artillery fire and infantry attacks. Finally, it must be noted that these concepts are extremely vulnerable to surprise attack and are not designed to restore territorial integrity by recapturing lost ground. Nevertheless, territorial defence units could be usefully added to manoeuvre forces, and NATO planners have already recognized that more effective use could be made of light infantry elements in the forward areas.

Preparing for conventional retaliatory offensive penetrations into Eastern Europe is not only beyond the reach of NATO's existing combat forces but would be destabilizing and politically unacceptable. The desirability and feasibility of conventional deep strikes with aircraft and long-range weapons must be assessed against clear objectives. NATO should improve its capabilities to disrupt Warsaw Pact reinforcement potential, but not at the cost of efforts to sustain initial defences at the levels of capability necessary to withstand and defeat Warsaw Pact first echelon forces. SACEUR's FOFA operational sub-concept is already being implemented, but disagreement persists among the Allies as to the weapons that should be used, the targets that should be attacked, and the most effective ranges for such operations. FOFA would probably be most effective if it concentrated on striking armoured combat units at ranges of less than 150 km from the line of contact, and on interdicting key choke-points and railway facilities at ranges out to 300 km rather than seeking to locate and attack mobile armoured concentrations at these longer ranges. In any case, the INF Treaty's ban on the development and deployment of ground-launched systems with ranges between 500 and 5 500 kilometres precludes the use of ground-launched missiles for very deep strikes.

Highest priority should be given to dealing with first echelon forces, to neutralizing Warsaw Pact artillery, and to the performance of offensive counter-air and active and passive air-defence operations. As one European military officer commented wryly, 'At the Battle of Little Big Horn, General Custer was not worried about the second echelon forces.'[5]

Before undertaking major new initiatives to strengthen conventional forces, Allied leaders will have to narrow their differences on defence policy and burden-sharing which have become increasingly polarized in recent years. Establishing such harmony will take time and a prudent transatlantic dialogue. In the process, West European governments will have to decide whether reducing NATO's reliance on nuclear weapons is affordable in resource terms or, conversely, whether failure to do so will both undermine domestic political support for defence and risk diminishing the American commitment to NATO. The revived WEU, while unlikely to achieve any dramatic success in promoting collaborative defence projects or creating a true European defence entity, may be useful in forging a European consensus on this question. In the end, it seems likely that NATO will manage to contrive a somewhat ragged consensus in support of a plan that combines conventional force modernization and arms control initiatives which together should enhance stability in Central Europe.

Notes and References

1 Introduction: The Search for a Stable Deterrent

1. See Sherri L. Wasserman, *The Neutron Bomb Controversy: A Study in Alliance Politics* (New York: Praeger, 1983).
2. McGeorge Bundy, George F. Kennan, Robert S. McNamara and Gerard Smith, 'Nuclear Weapons and the Atlantic Alliance', *Foreign Affairs*, 60, 3 (Spring 1982), p. 754; and Henry Kissinger, 'A Plan to Reshape NATO', *Time*, 5 March, 1984, pp. 20–4.
3. US Congress, Senate, Committee on Armed Services, 95th Congress, 1st Session, Committee Print, Report of Senators Sam Nunn and Dewey F. Bartlett, *NATO and the New Soviet Threat*, January 1977.
4. Bundy *et al.*, 'Nuclear Weapons'.
5. Karl Kaiser, Georg Leber, Alois Mertes and Franz-Josef Schulze, 'Nuclear Weapons and the Preservation of Peace', *Foreign Affairs*, 60, 4 (Summer 1982), pp. 1157–70. Another well-argued refutation of no-first-use, in this case by an American, is John J. Mearsheimer, 'Nuclear Weapons and Deterrence in Europe', *International Security*, 9, 3 (Winter 1984–5), pp. 16–46.
6. See Ivo H. Daalder, *The SDI Challenge to Europe* (Cambridge MA: Ballinger, 1987).
7. 'All Things Being Equal', *The Economist*, 18 April 1987.
8. US Congress, Senate, *Congressional Record*, Daily Edition, 18 June 1984, pp. S7451–9.
9. Bridget Bloom, 'U.S. "Railroading" Nato on High-Tech Weapons', *Financial Times* (London), 4 December 1983, p. 3.
10. See for comparison, the reports of the European Security Study ESECS, *Strengthening Conventional Deterrence in Europe* (New York: St Martin's, 1983) and *Strengthening Conventional Defense in Europe: A Program for the 1980s*, ESECS II (Boulder, Colorado and London: Westview 1985); and James R. Golden, Asa Clark and Bruce E. Arlinghaus (eds), *Conventional Deterrence* (Lexington, MA: D. C. Heath, 1984).

2 The Elusive Consensus

1. Stanley R. Sloan, *NATO's Future: Toward a New Transatlantic Bargain* (Washington, DC: National Defense University Press, 1985), p. 11.
2. Timothy P. Ireland, *Creating the Entangling Alliance* (Westport, CT: Greenwood Press, 1981), p. 207.
3. Francis A. Beer, *Integration and Disintegration in NATO* (Columbus: Ohio State University Press, 1969), p. 57.
4. Stanley R. Sloan, 'European Cooperation and the Future of NATO', *Survival*, 26, 6 (November/December 1984) p. 242.

5. A. W. DePorte, *Europe Between the Superpowers* (New Haven, CT: Yale University Press, 1979) pp. 157–65.
6. Samuel P. Huntington, *The Common Defense: Strategic Programs in National Politics* (New York: Columbia University Press, 1969), pp. 64–88.
7. US Department of State, *Foreign Relations of the United States, 1952–54*, vol. V, *Western European Security*, pp. 511–12.
8. Huntington, *The Common Defense*, pp. 99–106; and Timothy Ireland, 'Building NATO's Nuclear Posture: 1950–65', in Jeffrey D. Boutwell, Paul Doty and Gregory F. Treverton (eds), *The Nuclear Confrontation in Europe* (London: Croom Helm, 1985), pp. 10–11.
9. Robert E. Osgood, *NATO: The Entangling Alliance* (Chicago: University of Chicago Press, 1962), pp. 118–19; and Thomas W. Wolfe, *Soviet Power and Europe: 1945–1970* (Baltimore, MD: Johns Hopkins University Press, 1970), pp. 38–9, 141–3, 166.
10. William W. Kaufmann, 'The Requirements of Deterrence', in Kaufmann (ed.), *Military Policy and National Security* (Princeton, NJ: Princeton University Press, 1956), pp. 12–38.
11. David N. Schwartz, *NATO's Nuclear Dilemmas* (Washington, DC: Brookings, 1983), p. 51.
12. Robert S. McNamara, 'The Military Role of Nuclear Weapons', *Foreign Affairs*, 62, 1 (Fall 1983), pp. 63–4; and William W. Kaufmann, *The McNamara Strategy* (New York: Harper & Row, 1964), pp. 82–8.
13. McNamara, 'The Military Role', p. 63.
14. Catherine McArdle Kelleher, *Germany and the Politics of Nuclear Weapons* (New York: Columbia University Press), pp. 206–27.
15. Andrew J. Pierre, *Nuclear Politics: The British Experience with an Independent Strategic Force, 1939–1970* (Oxford: Oxford University Press, 1972), pp. 262–72.
16. General André Beaufre, *NATO and Europe* (New York: Vintage, 1966), pp. 67–71.
17. Schwartz, *NATO's Nuclear Dilemmas*, pp. 136–92.
18. 'The Future Tasks of the Alliance' (Harmel Report), December 1967, in UK, Secretary of State for Foreign and Commonwealth Affairs, *Selected Documents Relating to Problems of Security and Cooperation in Europe*, 1954–77, Cmnd 6932 (London: HMSO, 1977), Document 7, pp. 49–52.
19. 'Declaration on Mutual and Balanced Force Reductions, Reykajavik, June 1968', Ibid., Document 9, pp. 54–5.
20. Roger L. L. Facer, *Conventional Forces and the NATO Strategy of Flexible Response: Issues and Approaches* (Santa Monica, CA: RAND, 1985), p. 5.
21. Ibid., p. 33.
22. EUROGROUP Secretariat, *Western Defence: The European Role in NATO* (Brussels, 1984), p. 12.
23. Facer, *Conventional Forces*, pp. 33–5.
24. Steven L. Canby, 'NATO Muscle More Shadow Than Substance', *Foreign Policy*, no. 8, Fall 1972, pp. 38–49.

25. 'Defence Planning Committee, Final Communiqué, 18 May 1977', in *NATO Review*, July 1977, p. 26. The Nunn–Bartlett report received particular attention. See US Congress, Senate, Committee on Armed Services, 95th Congress, 1st Session, Committee Print, 'NATO and the New Soviet Threat', January 1977.

26. David Greenwood, 'NATO's Three Per Cent Solution', *Survival*, 23, 6 (November/December 1981), p. 253.

27. Ibid.

28. Robert W. Komer, *Treating NATO's Self Inflicted Wound*, RAND P-5092 (Santa Monica, CA: RAND, 1973).

29. US Department of Defense, *Annual Report to Congress: Fiscal Year 1980*, pp. 211–15.

30. US House, Committee on Government Operations, 97th Congress, 1st Session, Report 97–37, *The Implementation of the NATO Long Term Defense Program (LTDP)* 14 May, 1981, p. 8.

31. Ibid., p. 9.

32. US Department of Defense, Report to Congress by Secretary Caspar W. Weinberger, *Report on Allied Contributions to the Common Defense*, March 1984.

33. US Congress, House of Representatives, *Implementation of the NATO LTDP*, pp. 11.

34. David Watt, 'As Europe Saw It', *America and the World, 1986, Foreign Affairs*, 62, 3 (1986), pp. 521–32.

35. Gregory Flynn and Hans Rattinger, 'The Public and Atlantic Defense', in Flynn and Rattinger (eds), *The Public and Atlantic Defense* (Totowa, NJ: Rowman and Allanheld; London: Croom Helm, 1985), pp. 371–4.

36. US Congress, Senate, *Congressional Record*, Daily Edition, 18 June 1984, pp. S7451–9.

37. John E. Rielly, *American Public Opinion and U.S. Foreign Policy, 1987* (Chicago: Chicago Council on Foreign Relations, 1987), p. 21.

38. International Institute for Strategic Studies, *Military Balance 1985– 1986*, p. 198; *Military Balance 1986–1987*, p. 55; and *Strategic Survey 1986–87*, p. 104.

39. Ben J. Wattenberg and Karl Zinsmeister, 'The Birth Dearth: The Geopolitical Consequences', *Public Opinion*, December/January 1986, pp. 8–11.

40. Federal Republic of Germany, 'Report of the Commission for Long-Term Planning of the Armed Forces', 20 June 1982, and Ministry of Defence, *White Paper 1985: The Situation and Development of the Federal Armed Forces*, pp. 235–42; and IISS, *The Military Balance, 1983–84* (London: IISS, 1984), pp. 145–8.

41. Proceedings, Conference on Alternative Military Operational Concepts for NATO, John F. Kennedy School of Government, May 1986.

3 Opportunities for Cooperation

1. Alexander H. Cornell, 'Collaboration in Weapons and Equipment', *NATO Review*, October 1980, p. 18. See also Thomas Callaghan, *U.S.*

European Economic Cooperation in Military and Civil Technology (Washington, DC: Center for Strategic and International Studies, Georgetown University, 1975).

2. Keith Hartley, *NATO Arms Cooperation: A Study in Economics and Politics* (London: George Allen & Unwin, 1983), pp. 161–2.
3. John Stone, 'CNAD–Focal Point of Equipment Cooperation', *NATO Review*, 1 (1984), pp. 10–15; Johan Jorgen Holst, 'The Independent European Programme Group: Cooperation and Western Security', *NATO Review*, April 1981, pp. 6–9; and Jan van Houwelingen, 'The Independent European Programme Group (IEPG): The Way Ahead', *NATO Review*, 1984, pp. 17–21. The IEPG, which includes all NATO European governments, was established in 1976 to ensure effective use of European resources for research, development and procurement, expand standardization and interoperability, and expand the two-way street.
4. David Abshire, former US Ambassador to NATO, has been the most forceful American advocate of weapons collaboration. See David M. Abshire, 'Arms Cooperation in NATO', *Armed Forces Journal International*, December 1985, pp. 66–70. See also Benjamin F. Schemmer, 'Dep-SecDef Taft Weighs in Hard for Better NATO Armaments Cooperation', *Armed Forces Journal International*, January 1986, p. 30.
5. Michael Feazel, 'NATO Arms Cooperation Enters Critical Period', *Aviation Week and Space Technology*, 9 March 1987, pp. 79–81. See also Charles D. Odorizzi, 'Navy Wins 24 of 25 Projects Okayed for tional', January 1987, pp. 16–17.
6. Feazel, 'NATO Arms Cooperation', p. 81.
7. 'European Group Seeks Mutual Weapons Development Programs', *Aviation Week and Space Technology*, 9 March 1987, p. 82.
8. 'France, Germany Agree on New Helicopter', *Aviation Week and Space Technology*, 5 December 1983, p. 26; Robert Rudney, 'Franco–German Helicopter Back on Track', *Armed Forces Journal International*, May 1987, p. 27.
9. Paul Lewis, 'Europe's Fight over a Fighter', *New York Times*, 9 March 1985, p. 31.
10. Richard Evans, 'Arms Sales mean Big Business for France', *Christian Science Monitor*, 15 July 1985, p. 19. Despite President Mitterrand's declaration several years ago that France must stop selling arms to the Third World, France was the second largest arms exporter to that market in 1984, with a total sales volume of $9.1 billion.
11. David A. Brown, 'Future of European Fighter Uncertain as Governments Fail to Agree', *Aviation Week and Space Technology*, 3 June 1985, pp. 119–22.
12. 'Defense Ministries Review Future Fighter', *Aviation Week and Space Technology*, 9 July 1984, p. 22.
13. Jan Feldman, 'Collaborative Production of Defense Equipment within NATO', *Journal of Strategic Studies*, 7, 3 (September 1984), pp. 290–1.
14. Ibid., p. 292.

15. Benjamin F. Schemmer, 'Breaking Loose from NATO's Hang-Up over Numbers', *Armed Forces Journal International*, December 1985, pp. 71–2.
16. Giovanni de Briganti and Deborah G. Meyer, 'GTE/French System Wins MSE Award; Army says Decision Based on Cost', *Armed Forces Journal International*, December 1985, p. 22.
17. See speech by French Prime Minister Pierre Mauroy to the Institut des Hautes Études de la Défense Nationale, 'La Stratégie de la France', in *Défense Nationale*, November 1983, pp. 16–17.
18. Ivo H. Daalder and Lynn Page Whittaker, 'SDI's Implications for Europe: Strategy, Politics and Technology', in Stephen J. Flanagan and Fen Osler Hampson (eds), Securing Europe's Future (London: Croom Helm, 1985), p. 54.
19. See Guy de Jonquieres, 'EUREKA go-ahead for 62 Projects worth 2.1 bn', *The Financial Times*, 1 July 1986, p. 1; and a series of articles in *Le Monde Diplomatique*, August 1985, pp. 17–22.
20. 'Defence Planning Committee Communiqué', 5 December 1984, *NATO Review*, December 1984, p. 25.
21. David M. Abshire, 'NATO's Conventional Defense Improvement Effort: An Ongoing Imperative', *Washington Quarterly*, Spring 1987, p. 52–3.
22. 'Press Statement on Armaments Cooperation Issued by Foreign Ministers at the North Atlantic Council Meeting on 12th–13th December 1985', *NATO Review*, 6 (December 1985), pp. 23–4.
23. 'Defence Planning Committee Communiqué', *NATO Review*, 6 (December 1985), pp. 24–5. See also discussion in Chapter 7.
24. Abshire, 'NATO's Conventional Defense Improvement Effort', p. 55.
25. James M. Markham, 'U.S.–Soviet Missile Talks Pull Europeans Together', *New York Times*, 22 March 1987.
26. Sir Geoffrey Howe, 'The European Pillar', *Foreign Affairs*, 62, 2 (Winter 1984/5), pp. 330–43.
27. Werner Kern, 'Talks about 30-year old Treaty Reflect New Trend in Defense', *Saarbrucker Zeitung*, 27 June 1984, in *German Tribune*, 1140 (8 July 1984), p. 2.
28. Samuel F. Wells, Jr, 'The United States and European Defence Cooperation', *Survival*, 27, 4 (July/August 1985), pp. 161–3; and John Vinocur, 'Seven European Aides Meet to Bolster NATO Ties', *New York Times*, 13 June 1984, p. A5.
29. Elizabeth Pond, 'West European Moves to Strengthen NATO's European Pillar', *Christian Science Monitor*, 31 October 1984, p. 10; 'Europe Relaunches WEU as Forum for Defense', *Guardian*, 29 October 1984, p. 6; and 'If NATO became WETO', *The Economist*, 21 March 1987, pp. 61–3.
30. Kurt Becker, 'Bid to Create a New European Dimension to the Atlantic Pact', *Die Zeit*, 26 October 1984, in *German Tribune*, 1155 (4 November 1984), p. 2; and John Vinocur, 'West Europeans Restructure Arms Policy Panel', *New York Times*, 28 October 1984, p. 12.
31. Kurt Becker, 'West European Union Meets and Reaches Indecision',

Die Zeit, 26 April 1985, in *German Tribune*, 1127 (5 May 1985), p. 2.

32. William Wallace, 'European Defence Cooperation: The Reopening Debate', *Survival*, 26, 6 (November/December 1984), p. 251.
33. François Mitterrand, 'The Future of Europe', *World Today*, February 1987, pp. 40–2.
34. Markham, US–Soviet Missile Talks'.
35. Wells, 'The US and European Defence Co-operation', p. 165.
36. 'If NATO became WETO', p. 63.
37. Daalder, *The SDI Challenge*, p. 96.
38. Henry A. Kissinger, 'A Plan to Reshape NATO', *Time*, 5 March 1984, pp. 20–4.

4 France, Spain and Conventional Defence

1. Robert W. Komer, *Maritime Strategy or Coalition Defense?* (Lanham, MD: University Press of America, 1984), p. 83.
2. General Michel Fourquet, 'Emploi des différents systèmes de forces dans le cadre de la stratégie de dissuasion', *Revue de Défense Nationale*, May 1969, pp. 757–67.
3. David Yost, *France's Deterrent Posture and Security in Europe, Part I, Capabilities and Doctrine*, Adelphi Paper No. 194 (London: IISS, 1984), p. 9.
4. Kenneth Hunt, *NATO Without France: The Military Implications*, Adelphi Paper No. 32 (London: IISS, 1966), p. 7.
5. F. Roy Willis, *The French Paradox: Understanding Contemporary France* (Stanford, CA: Hoover Institution Press, 1982), p. 43.
6. Hunt, *NATO Without France*, p. 16.
7. 'Interview with General Bernard Rogers', *US News and World Report*, 15 June 1981, p. 25.
8. General Jeannou Lacaze, 'La Politique Militaire', *Défense Nationale*, November 1981, p. 10.
9. Yost, *France's Deterrent Posture*, p. 7.
10. IISS, 'French Defence Policy', *Strategic Survey*, 1976 (London: IISS, 1977), pp. 66–71.
11. IISS, *The Military Balance, 1974–75* (London: IISS, 1975), p. 21; and *The Military Balance, 1983–84* (London: IISS, 1983), pp. 32–3.
12. Yost, *France's Deterrent Posture*, p. 56.
13. Ambassade de France, Service de Presse, 'Reorganization of the French Army', 30 June 1983.
14. French Army Information Office, *Terre-Information*, 109 (June 1983). The new units are the 4th Airmobile Division and the 6th Light Armoured Division (which is an augmented 31st Brigade); the existing units are the 27th Alpine, 9th Naval Infantry and 11th Paratroop Divisions.
15. General René Imbot, 'The French Rapid Action Force', *NATO's Sixteen Nations*, Special 1/1983, pp. 34–9; and François Cailleteau 'La Force D'Action Rapide', *Etudes Polémologiques*, 3/1983 (December 1983), pp. 5–6.
16. Giovanni de Briganti, 'Fartel Exercise Reveals Limits of French Force

d'Action Rapide', *Armed Forces Journal International*, December 1985, p. 28.

17. General Georges Fricaud-Chagnaud, 'Origins, Capabilities and Significance of the Force d'Action Rapide', unpublished paper, Woodrow Wilson Center, Washington, DC, October 1984, p. 10.

18. Jonathan Marcus and Bruce George, MP, 'The Ambiguous Consensus: French Defence Policy under Mitterrand', *World Today*, October 1983, p. 373. Orders for 25 Mirage aircraft, artillery batteries and armoured vehicles were actually cancelled in 1983.

19. Yost, *France's Deterrent Posture*, pp. 56–9.

20. Robert Rudney, 'France, Boeing Sign $550-million Contract to Buy Three AWACS', *Armed Forces Journal International*, April 1987, p. 34.

21. 'The French are Ready to Cross the Rhine', *The Economist*, 13 July 1985, p. 44.

22. Deborah Caen, 'French Defense Equipment Budget Up, But Cuts Loom for Maintenance', *Armed Forces Journal International*, October 1986, p. 44.

23. Anne Chaussebourg, 'L'existence de la FAR n'implique pas un engagement automatique dans la "bataille de l'avant" déclare M. Hernu', *Le Monde*, 3 December 1984.

24. Yost, *France's Deterrent Posture*, p. 61.

25. Komer, *Maritime Strategy*, pp. 83–4; and Harold Brown, *Thinking About National Security: Defense and Foreign Policy in a Dangerous World* (Boulder, Colorado. Westview, 1983), pp. 103, 107.

26. David S. Yost, *France and Conventional Defense in Central Europe* (Boulder, Colorado: Westview, 1985), pp. 53–76.

27. Ibid., p. 57.

28. Ibid., p. 25.

29. John Vinocur, 'Paris–Bonn Military Ties: A Time for Reappraisal', *New York Times*, 20 October 1982, p. 11.

30. William Wallace, 'European Defence Cooperation: The Reopening Debate', *Survival*, 26, 6 (November/December 1984), p. 251.

31. Interviews with participants.

32. William Drozdiak, 'Bonn Seeks more Influence on French Nuclear Targeting', *Washington Post*, 20 April 1984, p. A19; and 'Schmidt Proposes Defense Merger by France, W. Germany', *Washington Post*, 29 June 1984, p. A21.

33. Kurt Becker, 'Europe Blueprints its own Security Policy', *Die Zeit*, 24 February 1984, in *German Tribune*, 1123 (4 March 1984), p. 1.

34. Drozdiak, 'Schmidt Proposes Defense Merger', p. A21; and 'Franco–German Ties Crucial, Says Schmidt', *General-Anzeiger Bonn*, 4 July 1984, in *German Tribune*, 1141 (15 July 1984), p. 2.

35. Pierre Lellouche, *L'Avenir de la Guerre* (Paris: Éditions Mazarine, 1985), pp. 279–88.

36. Ibid., p. 284.

37. François Heisbourg, 'Réalités et Illusions', *Le Monde*, 5 June 1985, p. 2.

38. 'Défense: Le Prix Des Ambitions', *L'Express*, 19 July 1985, pp. 13–

15; and 'Le PS Publié une Déclaration sur la Sécurité de l'Europe', *Le Monde*, 4 July 1985, p. 6; also in *Foreign Broadcast Information Service–Western Europe*, 8 July 1985, pp. K1–K2.

39. Pierre Lellouche, 'La Furia Française', *Le Point*, 660 (15 July 1985), pp. 24–5.
40. 'Bonn and Paris Plan Hot Line', *New York Times*, 23 August 1985, p. 15.
41. Federal Republic of Germany, Press and Information Office, 'Bonn Considering French Nuclear Protection', *The Week In Germany*, 11 October 1985, p. 2.
42. Paul Lewis, 'Paris–Bonn Military Accord is Reached', *New York Times*, 2 March 1986, p. 3; Paul Lewis, 'Franco–West German Brigade outside NATO is Planned', *New York Times*, 18 July 1987, p. 4; and Edward Cody, 'W. Europe Reevaluates its Defense', *Washington Post*, 13 July 1987, p. 13.
43. IISS, 'Europe: Security Uncertainties in NATO Countries', *Strategic Survey, 1986–1987*, pp. 102–3.
44. 'Interview with Eduardo Serra, Spain's Secretary of State for Defense', *Armed Forces Journal International*, August 1986, p. 88.
45. Ibid., p. 86.
46. Gregory F. Treverton, *Spain: Domestic Politics and Security Policy*, Adelphi Paper No. 204 (London: IISS, 1986), pp. 21–3.
47. Ibid., p. 23.
48. Ibid., p. 28.

5 The Impact of New Technologies: Evolutionary or Revolutionary?

1. ESECS, *Strengthening Conventional Deterrence in Europe* (New York: St Martin's, 1983), and *Strengthening Conventional Defense in Europe: A Program for the 1980s, ESECS II* (Boulder, Colorado, and London: Westview, 1985).
2. Sir Roy Mason, 'Lecture to the International Institute for Strategic Studies', 1983.
3. James A. Tegnelia, 'Emerging Technology for Conventional Deterrence', *International Defense Review*, 5 (1985), p. 646.
4. James W. Canan, 'Here Come the Superchips', *Air Force Magazine*, April 1984, pp. 48–54; and US Department of Defense, *Annual Report of the Secretary of Defense, Caspar W. Weinberger, to the Congress, Fiscal Year 1986* (Washington, DC: US Government Printing Office, 4 February 1985), p. 265.
5. Edgar Ulsamer, 'A Roadmap to Tomorrow's Tactical Airpower', *Air Force Magazine*, December 1983, p. 45.
6. US Department of Defense, Secretary Caspar W. Weinberger, *Annual Report to Congress, FY 1988*, 12 January 1987, pp. 200–1.
7. Tegnelia, 'Emerging Technology', p. 647.
8. John Burgess, 'Emerging Technologies and the Security of Western Europe', in Stephen J. Flanagan and Fen Osler Hampson (eds), *Securing Europe's Future*, pp. 64–84.

9. Gary Mitchelmore, 'The Big World of the Sentry', *Air Force Magazine*, April 1984, pp. 70–7; Brian Wanstall, 'Integrating the NATO AEW Force', *Interavia*, January 1984, p. 33; and Robert King, 'Nimrod Ax Changes UK Defense Plans', *Armed Forces Journal International*, February 1987, pp. 26–7.

10. Clarence A. Robinson, Jr, 'Surveillance Integration Pivotal in Israeli Successes', *Aviation Week and Space Technology*, 5 July 1982, pp. 16–17.

11. US Department of Defense, Secretary of Defense Caspar W. Weinberger, *Annual Report to the Congress, FY 1987*, 6 February 1986, p. 249.

12. Floyd D. Kennedy, Jr, 'The Radioelectronic Struggle: Soviet EW Doctrinal Development', *Signal*, December 1984, pp. 59–63; and Gerald Green, 'Soviet EW – Maskirova and REC', *National Defense*, April 1985, pp. 34–9.

13. Robinson, 'Surveillance Integration', p. 16.

14. J. P. Rapalski, 'DoD RPVs – Too Costly, Too Few, Too Late?', *Armed Forces Journal International*, February 1984, pp. 62–7; and Glenn W. Goodman, Jr, 'U.S. Military RPV Programs have taken Big Strides in 1986', *Armed Forces Journal International*, December 1986, pp. 66–70.

15. Lois M. Blake, 'Multiple Launch Rockets', *Journal of Defense and Diplomacy*, 1, 3 (June 1983) p. 51.

16. Edgar Ulsamer, 'Smart and Standing Off', *Air Force Magazine*, November 1983, p. 59.

17. Manfred Woerner and Peter Kurt Wurzbach, 'NATO's New "Conventional Option"', *Wall Street Journal*, 19 November 1982, p. 30.

18. Michael R. Gordon, 'Highly Touted Assault Breaker Weapon Caught up in Internal Pentagon Debate', *National Journal*, 22 October 1983, pp. 2152–6. Assault Breaker was not a single weapon but a number of components such as an airborne radar, tactical fusion processing centre, missile and payload, used in close coordination.

19. Ibid., p. 2155; and conversations with the author who obtained a copy of the report. See also US Congress, House of Representatives, Committee on Armed Services, 98th Congress, 1st Session, Committee Print No. 13, Staff Study, *Improved Conventional Force Capability: Raising the Nuclear Threshold*, 1984, pp. 7–9.

20. Ulsamer, 'Smart and Standing Off', p. 61.

21. ESECS, *Strengthening Conventional Deterrence*, p. 238.

22. ESECS, *Strengthening Conventional Defense*, pp. 52–65.

23. 'Patriot-Roland Deal wrapped up', *Armed Forces Journal International*, August 1984, p. 34.

24. IISS, *The Military Balance, 1985–1986* (London: IISS, 1985), p. 162.

25. US Departments of Defense and State, *Soviet Strategic Defense Programs*, October 1985, p. 20; and Robert M. Gates and Lawrence K. Gershwin, Central Intelligence Agency, 'Soviet Strategic Force Developments', Testimony before a Joint Session of the US Senate Armed Services and Defense Appropriations Committees, 26 June 1985, p. 5 (mimeo).

26. For a discussion of SDI work relevant to the ATBM question, see Gregory H. Canavan, *Theater Applications of Strategic Defense Concepts* (Los Alamos, N.M.: Los Alamos National Laboratory, June 1985), LA-UR-85-2117 (P/AC: 85–149).
27. Clarence A. Robinson, Jr, 'U.S. Develops Antitactical Weapon for Europe Role', *Aviation Week and Space Technology*, 9 April 1984, pp. 46–9.
28. Michael Feazel, 'German Minister Proposes Initiative to Improve European Defenses', *Aviation Week and Space Technology*, 9 December 1985, pp. 19–20; and Manfred Woerner, 'A Missile Defense for NATO Europe', *Strategic Review*, 14 (Winter 1986), pp. 13–19.
29. 'Air Defense Laser', *International Defense Review*, 12/1985, p. 2036.
30. Blake, 'Multiple Launch Rockets', p. 53.
31. Alan Dodd Frank, 'Learning the Hard Way', *Forbes*, 6 June 1983, pp. 41–2.
32. William J. Perry and Cynthia A. Roberts, 'Winning Through Sophistication: How to Meet the Soviet Challenge', *Technology Review*, July 1982, pp. 27–35.
33. See John J. Mearsheimer, 'Precision-Guided Munitions and Conventional Deterrence', *Survival*, 21, 2 (March/April 1979), pp. 68–76; Daniel Gouré and Gordon McCormick, 'PGM No Panacea', *Survival*, 22, 1 (January/February 1980), pp. 15–19 and John J. Mearsheimer 'Rejoinder', *Survival*, 22, 1 (January/February 1980), pp. 20–22; and Michael L. Brown and Thomas J. Leney, 'Conventional Defense: Technology, Doctrine and Force Structure', in J. R. Golden, A. Clark, and B. E. Arlinghaus, *Conventional Deterrence* (Lexington, MA: Heath, 1984), pp. 163–76.
34. Benjamin F. Schemmer, 'Interview With Philip A. Karber', *Armed Forces Journal International*, May 1987, pp. 42–60; and Vernon A. Guidry, Jr, 'Army Modifies Missile, Seeks Edge on Armor', *Baltimore Sun*, 19 July 1987, p. 1.
35. ESECS, *Strengthening Conventional Defense*, pp. 133–5.
36. B. Bloom, 'U.S. "Railroading" NATO on High-Tech Weapons', *Financial Times* (London), 4 December 1983, p. 3.
37. David A. Brown, 'NATO Selects Emerging Technologies', *Aviation Week and Space Technology*, 16 April 1984, pp. 28–9.

6 Changes in Concepts and Tactics

1. General Bernard W. Rogers, 'Greater Flexibility for NATO's Flexible Response', *Strategic Review*, 11 (Spring 1983), pp. 11–19.
2. ESECS, *Strengthening Conventional Deterrence* and *Strengthening Conventional Defence*.
3. Philip A. Karber, 'To Lose an Arms Race: The Competition in Conventional Forces Deployed in Central Europe, 1965–1980', in Uwe Nerlich (ed.), *The Soviet Asset: Military Power in the Competition over Europe* (Cambridge, MA: Ballinger, 1983), pp. 31–88.
4. Boyd D. Sutton, John R. Landry, Malcolm B. Armstrong, Howell M. Estes III and Wesley K. Clark, 'Deep-Attack Concepts and the

Estes III and Wesley K. Clark, 'Deep-Attack Concepts and the Defence of Central Europe', *Survival*, 26, 2 (March/April 1984), pp. 50–70.

5. US Army, *Field Manual (FM) 100-5: Operations* (Washington, DC: Department of the Army, 1982): hereinafter cited as FM 100-5.

6. John J. Mearsheimer, 'The Military Reform Movement: A Critical Assessment', *Orbis*, 27, 2 (Summer 1983), p. 291.

7. Colonel William G. Hanne, 'AirLand Battle – Doctrine not Dogma', *International Defense Review*, August 1983, pp. 1035–36.

8. *FM 100-5*, pp. 7-12–7-17. The Army manual actually refers to 'Deep Battle' in the Corps Commander's area of influence, which is defined as extending far enough beyond the FLOT to 'engage enemy forces which can join or support the main battle within 72 hours'. In general this implies a geographic coverage out to 100–150 km.

9. Huba Wass de Czege, 'Army Doctrinal Reform', in Asa A. Clark, Peter W. Chiarelli, Jeffrey S. McKitrick and James W. Reed, (eds), *The Defense Reform Debate: Issues and Analysis* (Baltimore, MD: Johns Hopkins University Press, 1984), p. 108.

10. Michael Gordon, ' "ET" Weapons To Beef Up NATO Forces Raise Technical and Political Doubts', *National Journal*, 19 February 1983, p. 369.

11. Sutton *et al.*, 'Deep-Attack Concepts', pp. 61–2.

12. US Department of the Army and Department of the Air Force, 'Memorandum of Agreement on U.S. Army–U.S. Air Force Joint Force Development Process', 22 May 1984 (mimeo).

13. Ibid. See also Edgar Ulsamer, 'The AirLand Agreement', *Air Force Magazine*, July 1984, pp. 16–19; and Benjamin F. Schemmer, ' "Historic" Army/USAF Agreement to Reduce Roles and Missions Overlap', *Armed Forces Journal International*, July 1984, pp. 19–26.

14. General Bernard W. Rogers, 'Follow-On Forces Attack (FOFA): Myths and Realities', *NATO Review*, December 1984, p. 1.

15. US, Congress, Senate, Committee on Armed Services, 98th Congress, 1st Session, Senate Hearing 98–49, Part 5, *Department of Defense Authorization for Appropriations for Fiscal Year 1984, Strategic and Theater Nuclear Forces*, March–May 1983, p. 2406.

16. Ibid., p. 2407.

17. Interviews, Bonn, January 1984.

18. See Christopher Donnelly, 'The Development of the Soviet Concept of Echeloning', *NATO Review*, December 1984, p. 9; and 'The Soviet Operational Manoeuvre Group: A New Challenge for NATO', *Military Review*, March 1983, pp. 43–60. See also Benjamin Schemmer, 'U.S. Unveils Multi-Source Evidence of New Threats in German Forum', *Armed Forces Journal International*, August 1984, p. 51.

19. Rogers, 'Follow-On Forces Attack', p. 4.

20. Ibid., p. 8.

21. Sutton *et al.*, 'Deep-Attack Concepts', pp. 60–1.

22. Rogers, 'Follow-On Forces Attack', p. 7.

23. Michael R. Gordon, 'Highly Touted Assault Breaker Weapon Caught up in Internal Pentagon Debate', *National Journal*, 22 October 1983, pp. 2154–5.

24. Clarence A. Robinson, Jr, 'U.S. Develops Antitactical Weapon for Europe Role', *Aviation Week and Space Technology*, 9 April 1984, p. 47.
25. US, Congress, House of Representatives, Committee on Armed Services, 98th Congress, 1st Session, Committee Print No. 13, Staff Study, *Improving Conventional Force Capability: Raising the Nuclear Threshold*, 1983.
26. See 'Denmark Refuses Non-Nuclear Cruise', *Sunday Times* (London), 18 December 1983; and Nils Peter Gleditsch, 'The Freeze in Norway', *Bulletin of the Atomic Scientists*, November 1983, pp. 32–4.
27. Steven Canby, 'Military Reform and the Art of War', *Survival*, 25, 3 (May/June 1983), pp. 120–3; and Steven Canby, *The Alliance and Europe: Part IV, Military Doctrine and Technology*, Adelphi Paper No. 109 (London: IISS, 1975), p. 1.
28. For a description of the roles and missions of the *Bundesmarine*, see Federal Republic of Germany, Minister of Defence, *White Paper 1985: The Situation and the Development of the Federal Armed Forces*, 19 June 1985, pp. 211–20.
29. Ingemar Dorfer, 'Technological Development and Force Structure Within the Atlantic Alliance: Prospects for Rationalization and the Division of Labour', in *New Technology and Western Security Policy*, Adelphi Paper No. 197 (London: IISS, 1985), p. 44.
30. Ibid.
31. Steven Canby and Ingemar Dorfer, 'More Troops, Fewer Missiles', *Foreign Policy*, 53 (Winter 1983/4), pp. 10–13.
32. Canby, 'Military Reform', p. 121; and 'Territorial Defense in Central Europe', *Armed Forces and Society*, 7, 1 (Fall 1980), pp. 51–67. For a discussion of the new West German army structure, see FRG, *White Paper 1985*, pp. 195–6.
33. Canby, 'Military Reform', p. 122.

7 A Strategic Concept for Conventional Defence

1. 'DPC Communiqué, 22 May 1985', in *NATO Review*, 3 (June 1985), pp. 31–2; and 'DPC Communiqué, 22 May 1986', in *NATO Review*, 3 (June 1986), pp. 30–1. See also IISS, 'Conventional Defence Improvements in NATO', *Strategic Survey 1985–1986*, (London: IISS, 1986), pp. 38–40.
2. John J. Mearsheimer, *Conventional Deterrence* (Ithaca, NY: Cornell University Press, 1983), p. 165.
3. US Department of Defense, Report of the Secretary of Defense, Caspar W. Weinberger, to the US Congress, *Improving NATO's Conventional Capabilities*, June 1984, p. vii. (Declassified version released under the Freedom of Information Act.)
4. General Bernard W. Rogers, 'NATO's Strategy: An Undervalued Currency', in IISS, *Power and Policy: Doctrine, the Alliance and Arms Control*, Adelphi Paper No. 205 (London: IISS, 1986), p. 5.
5. D. Yost, *France's Deterrent Posture and Security in Europe, Part I*,

Capabilities and Doctrine, Adelphi Paper No. 194 (London: IISS, 1984), pp. 62–3.
6. Rogers, 'NATO's Strategy', p. 7.
7. Federal Republic of Germany, Ministry of Defence, *The Situation and Development of the Federal Armed Forces, White Paper 1985*, 19 June 1985, pp. 391–2.
8. William W. Kaufmann, 'Nonnuclear Deterrence', in John D. Steinbruner and Leon V. Sigal (eds), *Alliance Security and the No-First-Use Question* (Washington, DC: Brookings, 1983), p. 164.
9. Interviews, Bonn, January 1984.
10. Mearsheimer, *Conventional Deterrence*, p. 165.
11. Kaufmann, 'Nonnuclear Deterrence', p. 71.
12. Ibid., p. 73.
13. Ibid., p. 73.
14. Ibid., pp. 63–75; see also US Department of Defense, Assistant Secretary of Defense Program Analysis and Evaluation, 'NATO Center Region Military Balance Study, 1978–1981', July 1979, pp. I–50–52. (Declassified 13 July 1985.)
15. Samuel P. Huntington, 'Conventional Deterrence and Conventional Retaliation in Europe', *International Security*, 8, 3 (Winter 1983–4), pp. 32–56.
16. Ibid., p. 40.
17. Ibid., pp. 48–9.
18. Keith A. Dunn and William O. Staudenmaier, 'A NATO Conventional Retaliatory Strategy: Its Strategic and Force Structure Implications', in Dunn and Staudenmaier (eds), *Military Strategy in Transition: Defense and Deterrence in the 1980s* (Boulder, Colorado, and London: Westview, 1984), p. 202.
19. Ibid., pp. 202–3.
20. Ibid., p. 207.
21. Huntington, 'Conventional Deterrence', p. 55.
22. Eckhard Lubkemeier, 'Deterrence, Detente and Defense in Europe: How not to Reform NATO's Strategy', paper of the Research Institute of the Friedrich Ebert Stiftung, May 1984, p. 10.
23. SPD Germany, 'Peace and Security', Motion for the Party Conference, Nuremberg, 25–9 August 1986, p. 5.
24. Quoted in Adam Roberts, *Nations in Arms*, 2nd edn (New York: St Martins, 1986), p. 258.
25. SPD, 'Peace and Security', p. 7.
26. Karsten Voigt, 'Interim Report of the Sub-Committee on Conventional Defence: New Strategies and Concepts', *North Atlantic Assembly*, November 1986, pp. 20–7.
27. Andreas von Bülow, 'Defensive Engagement: An Alternative Strategy for NATO', in Andrew Pierre (ed.), *The Conventional Defense of Europe* (New York: Council on Foreign Relations, 1986), pp. 112–51.
28. Egbert Boeker and Lutz Unterseher, 'Emphasizing Defense', in Frank Barnaby and Marlies ter Borg (eds), *Emerging Technologies and Military Doctrine* (New York: St Martin's, 1986), p. 89.

29. Dr Albrecht A. C. von Müller of the Max Planck Institute, Starnberg, West Germany, has provided a lucid exposition of this argument in his unpublished paper, 'Integrated Forward Defense: Outlines of a Modified Conventional Defense for Central Europe', 1985, p. 8, and 'Pugwash Report on Conventional Defense', *Pugwash Newsletter*, July 1986, p. 114.
30. Ibid.
31. IISS, *Strategic Survey*, 1986–7, p. 102.
32. Boeker and Unterseher, 'Emphasizing Defense', p. 91.
33. Hew Strachan, 'Conventional Defence in Europe', *International Affairs* (London), 61, 1 (Winter 1984/5), p. 31, and Boeker and Unterseher 'Emphasizing Defense', pp. 95–97.
34. Jochen Löser, 'The Security Policy Options for Non-Communist Europe', *Armada International*, 2, 82 (March/April), pp. 66–75.
35. Von Müller, 'Integrated Forward Defense', pp. 19–25.
36. Norbert Hannig, 'Can Western Europe be Defended by Conventional Means?', *International Defense Review*, 1 (1979), pp. 27–34.
37. Boeker and Unterseher, 'Emphasizing Defence', pp. 102–3.
38. Frank Barnaby and Stan Windass, *What is Just Defence?* (Oxford: Just Defence, 1983).
39. Report of the Alternative Defense Commission, *Defence Without the Bomb* (New York: Taylor & Francis, 1983), pp. 8–11, 249–79.
40. SPD, 'Peace and Security'.
41. Voigt, 'Interim Report', p. 28.
42. Government of the Netherlands, Ministry of Defence, Memorandum by the Minister of Defence, Jacob de Ruiter, 'Reinforcement of the Conventional Defence and "Emerging Technologies",' 26 June 1985, paragraph 26.
43. H. W. Hofmann, R. K. Huber and K. Steiger, 'On Reactive Defense Options', in Reiner K. Huber (ed.), *Modelling and Analysis of Conventional Defense in Europe: Assessment of Improvement Options* (New York and London: Plenum Press, 1986), p. 138.
44. Wilhelm Nolte, 'America's Nuclear Deterrence', unpublished paper delivered at the International Studies Association Meeting, Washington, DC, April 1987.

8 Twin Challenges

1. M. Proskurin, 'The Aggressive Nature of the Rogers Plan', *Krasnaya Zvezda*, 29 October 1983, p. 5. Reprinted in *Foreign Broadcast Information Service – Soviet Union*, 4 November 1983, p. C1.
2. Ben J. Wattenberg and Karl Zinsmeister, 'The Birth Dearth: The Geopolitical Consequences', *Public Opinion*, December/January 1986, p. 9.
3. William E. Odom, 'Soviet Force Posture: Dilemmas and Directions', *Problems of Communism*, July–August 1985, pp. 6–14.
   Mary C. Fitzgerald, 'Marshal Ogarkov on the Modern Theater Operation', *Naval War College Review*, Autumn 1986, pp. 6–22.
5. Stephen M. Meyer, *Soviet Theatre Nuclear Forces, Part II: Capabilities*

and Implications, Adelphi Paper No. 188 (London: IISS, 1984), pp. 24–53.

6.   US Department of Defense, *Soviet Military Power 1985* (Washington, DC: US Government Printing Office, 1985), pp. 67–8.

7.   F. D. Sverdlov, 'The Tactical Maneuver', in *Joint Periodical Reference Service*, LI 11388, 16 June 1983, p. 84.

8.   'Ogarkov Interview with Krasnaya Zvezda', 9 May 1984, reprinted in *Survival*, 26, 4 (July/August 1984), p. 186.

9.   Sverdlov, 'The Tactical Maneuver', pp. 29–30.

10.   John G. Hines and Phillip A. Peterson, 'The Soviet Conventional Offensive in Europe', *Military Review*, April 1984, pp. 3–29.

11.   See Clarence A. Robinson, Jr, 'U.S. Develops Antitactical Weapon for Europe Role', *Aviation Week and Space Technology*, 9 April 1984, p. 49.

12.   'Statement on the Nuclear Threat by the Warsaw Treaty Member States', *Pravda*, 24 October 1985, pp. 1–2, in *The Current Digest of the Soviet Press*, 37, 43 (20 November 1985), pp. 4–5. See also statement of Marshal Kulikov, 'For the Sake of Peace on Earth', *Izvestia*, 8 May 1984, pp. 1–2, in *The Current Digest of the Soviet Press*, 36, 19 (31 May 1984), pp. 11–12.

13.   Extracts from the Budapest Appeal can be found in *NATO's Sixteen Nations*, July 1986, pp. 84–5. The Communique of the Berlin Political Consultative Committee Meeting of May 29, 1987 is reproduced in Foreign Broadcast Information Service (FBIS), *Eastern Europe Report*, 1 June 1987, pp. AA7–AA18.

14.   *Whence the Threat to Peace?*, 4th edn (Moscow: Novosti Press Agency Publishing House, 1987), p. 75ff.

15.   See 'Halifax Statement on Conventional Arms Control', in *NATO's Sixteen Nations*, July 1986, pp. 84–5.

16.   See James A. Thompson and Nanette C. Gantz, 'Conventional Arms Control Revisited: Objectives in the New Phase', Paper presented at the Conference on 'Arms Control and Conventional Defence in Europe', Berlin, September 1987, p. 10; and Stephen J. Flanagan and Andrew Hamilton, 'Alarm Bells and Zonal Force Constraints: An Alternative Approach to Conventional Stability in Europe', in John Borawski (ed.), *A Better Peace?* (London and Washington: Pergamon/Brasseys, Forthcoming).

17.   See Jay Kosminsky, 'European Nuclear Security: Beyond Current Dilemmas', in Flanagan and Hampson, *Securing Europe's Future*, pp. 15–16.

18.   Jonathan Dean, *Watershed Europe* (Lexington, MA: Heath, 1987), p. 155.

9   Conclusion: Improving NATO's Conventional Defences

1.   See Philip A. Karber, 'To Lose an Arms Race: The Competition in Conventional Forces Deployed in Central Europe, 1965–1980', in Uwe Nerlich (ed.), *The Soviet Asset: Military Power in the Competition over Europe* (Cambridge, MA: Ballinger, 1983), pp. 31–88; and NATO

Information Service, *NATO and the Warsaw Pact: Force Comparisons*, 1984, and the more favourable assessments of NATO's situation in the Central Region provided by Barry Posen, 'Measuring the European Conventional Balance: Coping with Complexity in Threat Assessment', *International Security* 9, 3 (Winter 1984–5), pp. 47–89; John Mearsheimer, 'Why the Soviets Can't Win Quickly in Central Europe', *International Security*, 7, 1 (Summer 1982), pp. 3–39; Anthony Cordesman, 'The NATO Central Region and the Balance of Uncertainty', *Armed Forces Journal International*, July 1983, pp. 18–58.

2. See John C. F. Tillson IV, 'The Forward Defense of Europe', *Military Review*, 61 (May 1981), pp. 66–76; Elizabeth Pond, 'West Germany says "nein" to liquid mines', *Christian Science Monitor*, 27 August 1984, p. 9; and Wayne Biddle, 'US Describes Pipeline Defense Against Tanks', *International Herald Tribune*, 25 August 1984, p. 1.

3. See John Mearsheimer, 'Nuclear Weapons and Deterrence in Europe', *International Security* Vol. 9 No. 3 (Winter 1984–5), pp. 19–46.

4. Jay Kominsky, 'European Nuclear Security: Beyond Current Dilemmas', in S. J. Flanagan and F. O. Hampson (eds), *Securing Europe's Future* (London: Croom Helm, 1985), pp. 15–16.

5. Interview.

Index